Vanished Origins

by

W. M. House

Published by Archean Enterprises, LLC

Vanished Origins

Published by: Archean Enterprises, LLC

16200 SW Pacific Highway
Ste H #2055
Tigard OR 97224
United States

Cover Design by ArcheanArt

Author: W. M. House

Art: AI Generated Images Modified by ArcheanArt

ISBN:

Paperback: 978-1-958457-12-2

E-Book: 978-1-958457-11-5

....

Beginnings unknown

Worlds lost in the mists of time

Paths to the present

.

.

Evolution is a force relentlessly pushing life into the future,
regardless of whether humanity wants to go.

Prologue

The human mind always seeks to understand what it sees and what it doesn't see.

This innate curiosity forms the origins of science and philosophy.

Homo Sapiens & Knowledge

We should not think of evolution as a series of discrete steps marking the rise and fall of species. Evolution is a fundamental force of life that binds together our biosphere. It is always there, forcing change and shaping the future. Evolution is the essential process by which life has spread across the surface of our planet. It allows life to test and keep what works and eliminate what doesn't. Through evolution, the threads of all life on Earth bind together into a symbiotic fabric we think of as Gaia. No species, Homo sapiens included, can exist on its own, separate from the tapestry of our biosphere. All life is interconnected.

Homo sapiens are the end product of four billion years of evolution.

We are like all other species in many ways. Our members live and die, passing their genetic legacy to future generations. We are omnivores huddling in groups for survival, eating other life forms for sustenance. But in other ways, we are very different. No other species has gained mastery of the planet like humans. We transport ourselves on land, air, and sea via mechanical inventions, and we can traverse Earth's great oceans in hours. How is it that we have accomplished what the countless millions of species that have gone before us failed to achieve?

The human mind traverses the arc of time, remembering the past and envisioning the future. We are a species uniquely attuned to the fundamental nature of information. Our minds effectively collect information and tie it into working models that we project into the future. We see what is and transform it into what might be. Other animals process information for day-to-day survival. We, however, take it to another level and use information to mold a more survivable future. But this is not the key characteristic that makes Homo sapiens unique.

The genius of the human species is its ability to preserve new learning and pass it on to the next generation. Henry Ford could not have manufactured the first Model T automobiles without the metallurgical and engineering expertise that humanity had accumulated over centuries before his arrival. Likewise, Bill Gates would not have revolutionized the personal computer without the vast array of technology passed on to him by our forebears. We store useful information in oral, written, and electronic formats for future reference, and we collectively refer to it as knowledge.

In the vast 13.8 billion-year history of the universe, human beings are less than a period at the end of a novel. If we shrink the universe's existence to 24 hours, then 0.13 seconds before midnight at the end of that day, humans gained enough existential awareness for civilization to take off. Interestingly, when scientists measure the blink of an eye, they estimate its duration to be approximately 100 to 150 milliseconds or around 0.13 seconds. So, during the universe's long history, in this last blink of an eye, the Torah, Bible, and Quran were written, huge metropolitan cities arose from the ground, men walked on the moon, and humanity discovered the quantum weirdness at the heart of our universe.

Yet these achievements were only possible because each generation could build upon humanity's cumulative knowledge base, adding to it for the benefit of the next generation.

Knowledge comes in discrete packets, not in a steady flow. It consists of brief spurts of genius and insight that arrive periodically and unexpectedly. The process of accumulating and incorporating new information into the basket of human learning is slow. Einstein, Leonardo da Vinci, and the unknown inventor of the Clovis Point are three examples among millions of brilliant minds whose observations form our civilization's technological and philosophical foundations.

The episodic and unpredictable appearance of new and unique knowledge is a blessing and a curse. The blessing is that it happens. The curse is that the true genius needed to move humanity forward in big leaps happens infrequently. Humans are excellent at taking new ideas and utilizing them to gain wealth and power, but they are poor at extrapolating what is known to find the unknown jewels of wisdom needed to move society forward.

Unfortunately for Galileo, when individuals have brilliant insights, society often beats them down to squash their ideas and prevent their new knowledge from being passed on to future generations. New knowledge displaces existing conventions, disrupting the status quo. Those with the most to lose often resist the changes that result from new ideas, and the natural reaction of people with accumulated wealth and power is to resist change if it threatens their status.

By and large, we are a risk-averse species, preferring to cling to a familiar past rather than embracing an uncertain future. The best vision of the future that many people can muster is a return to the past rather than a vision of moving forward to a new, promising future. They ignore the fact that time, as we understand it, only moves in one direction: forward. We cannot return to the past; the future is all we have.

Ironically, as knowledge accumulates and the information we have at our fingertips increases exponentially, our ignorance grows even faster. For example, we live in a universe dominated by dark matter. Understanding the presence of dark matter in our cosmos is a testament to the genius of the human mind, but it lays bare the fact that we have no idea what dark matter is. While the discovery of dark matter increased our knowledge base, it also exposed a vast ocean of ignorance. Before the discovery of dark matter, we assumed that our journey toward understanding the visible and detectable universe was making great progress. We were rapidly increasing our knowledge and decreasing our ignorance. The new knowledge that our universe is primarily composed of a mysterious substance that we can't see or directly measure has immensely expanded our collective ignorance; every answer creates a plethora of questions and unknowns.

Overall, Homo sapiens are a marvelously ingenious species, but they can also be cruel and vindictive, spawning wars where millions of men, women, and children needlessly die. Human genius can be used to mold a brilliant future or send our species into the void of extinction, taking much of life on Earth with us. Perhaps we fail to understand the context of our existence. Homo sapiens are the first species in the planet's long history to deliberately reshape the biosphere, rapidly alter the climate, and significantly impact the future of all life on Earth. Ironically, in our quest to dominate the planet, we may rob ourselves of our future. We should contemplate evolution's primary lesson: change is inevitable, and species extinction is the rule, not the exception.

....

A Quantum World

The year was 1935 when Einstein, Podolsky, and Rosen published their famous paper outlining the "EPR" Paradox. One of the peculiarities noted in the paper was quantum entanglement, referring to the observation that once two sub-atomic particles interact, they can become connected in ways we can't fully explain. Einstein referred to it in German as "*spukhafte fernwerkung*,"—spooky action at a distance. If you observe the state of one entangled particle, then you know the state of the other, whether they are separated by two centimeters or 200 light years. If the state of one changes, then the state of the other changes.

Quantum observations are quirky. A mathematical probability wave defines a photon until we observe it. The photon could be in many different locations, and the wave function determines the probability it is in a particular position. Only when we observe the photon does it manifest itself as an object at a specific point in space and time. Is it everywhere until we see it, or is it nowhere until we measure it? It may follow all possible probability paths simultaneously in multiple universes or only display particle behavior in the reference frame of our universe. These are all thoughts proposed by quantum physicists and a reflection of our shallow understanding concerning the nature of the quantum world.

Quantum observations about photons do raise interesting questions about the underlying nature of reality. Niels Bohr once said, "*Everything we call real is made of things that cannot be regarded as real.*" Bohr laid the scientific foundations for quantum theory and our understanding of atomic structure. Yet, he sometimes harbored an almost mystical view of reality and repeatedly questioned physicists' ability to independently observe objects in nature.

How can you independently observe an object within a system to which you are inextricably bound? You can't position yourself outside of the system. You can only observe the object from your position within the system. Your understanding is the outcome of how you and the object interact, not a description of the object in an independent state. Reality becomes subjective and depends on where in the system you are when you make an observation. Philosophers will recognize subjective reality as an existential concept since we are all constrained to defining the world around us through the lens of our individual experiences.

Another factor in the quantum enigma is that once the location of a photon is precisely determined, the probability wave collapses. The fuzzy spatial nature of the photon transforms into a point. A single photon fired through a single small slit can potentially strike the wall behind the slit at multiple locations. This uncertainty does not result from a lack of knowledge about speed and direction, as per the analysis of motion in classical physics. This uncertainty reflects the photon's wave-like behavior. The wave equation can only describe the probability of the photon striking at any single position. It can't predict the actual location of the strike.

If the photon strikes at point A, then the probability that it will also strike at point B goes to zero. The probability wave collapses. How does the wave know when to collapse? Why can't it produce action at multiple points, and what happens to the "information" field represented by the wave? Einstein's view of quantum entanglement as *spukhafte fernwerkung* also applies to the collapse of a probability wave.

Philosophically (not mathematically), the collapse of a probability wave is similar to the concept of time espoused by the early Christian philosopher Saint Augustine. He argued that the past is gone, and we retain it only as an imperfect, remnant memory. The future is yet to arrive and lacks reality. Thus, the present is the only place where we exist as an active force in the universe. The present, however, is elusive and slips away as soon as it arrives. It can only exist at the non-dimensional interface between the past and future and has no independent state of being.

The future is unknown and thus consists of an array of probabilities concerning events that might or might not occur. As time moves forward and the present progresses to a point in the future, the array of possible future events collapses. A single event finally occurs when the future becomes the present. Similarly, once the photon interacts with the universe, it becomes observable, and the probability wave collapses.

The Unknown

We do not know whether quantum entanglement is the result of an unknown dimension intersecting with the space-time continuum or if entanglement is a fundamental property of the universe, and space-time dimensions, as we understand them, are emergent properties springing from quantum entanglement.

However, the story that follows is not intended to answer these questions about the fundamental nature of our universe. Instead, it acknowledges the complex and often conflicted nature of the human mind and ponders what that means for the future. What was the evolutionary path leading to a species whose mind can conceive of the quantum mysteries at the heart of our universe?

As stated previously, we are the end product of four billion years of evolution, a species capable of engineering its own demise. The following chapters look backward in time to our vanished and forgotten origins and attempt to place context around the development of life on Earth. The hope is that by better understanding humanity's symbiotic relationship with the planet's biosphere, we might appreciate the responsibility to the planet that our species inherited by becoming the most dominant agent of change ever to arise from biological evolution on planet Earth.

Water

Water nourishes all of life without trying.

Water is content to fall from the sky, or flow through the low places, staying close to the ground.

Water flows as it will, reducing the mountains to hills.

It causes life to spring from the fecund soil.

Water is the sine qua non of life, that which we cannot do without; it is the magic elixir making our reality possible. Earth, the third planet from Sol, precariously balanced between searing heat and the frigid cold of deep space, occupies the sweet spot for life as we know it.

The universe hummed along for nine billion years before events in a far corner of the Milky Way Galaxy conspired to create a moderate-sized planet in the orbit of a relatively new star. Thus far, it is the only planet in the universe known to sustain biological life. Granted, this claim to fame is more of a statement of ignorance than knowledge since our understanding of the universe is meager at best. In a universe estimated to contain 200 billion trillion stars, it seems inevitable that trillions of planets circling these stars will occupy an orbit within the Goldilocks zone, where heat and cold balance out, allowing liquid water to pool in the topographic lows, forming oceans, lakes, and rivers. From a statistical perspective, the development of life on Earth is more of a template for how life evolves throughout the universe than a unique occurrence.

The earliest version of Earth was a vision of hell. Geologists refer to the first 500 million years of Earth's history as the Hadean Eon in honor of Hades, the Greek god of the underworld. Cosmic debris collapsed into a proto-earth gravity well, generating heat as bits and pieces of metal, stone, and molecular dust packed together into progressively denser configurations. Heat and molten fury dominated the planet. There was no niche for life to exploit. The early Earth needed to cool before the water in its atmosphere could condense into liquid, forming our first oceans. The jury is out on the exact date this happened, but the most common view is that it occurred sometime before 3.8 billion years ago.

Many proposed dates in Earth's history, including the first appearance of liquid water, rely on geological evidence. For example, the 3.8 billion-year date just mentioned rests on the presence of pillow basalts in the Isua Greenstone Belt of southwestern Greenland. These ancient rocks formed approximately 3.8 billion years ago, and since pillow basalts typically form below water, we can safely infer that the Earth had large bodies of water when these basalts erupted on an ancient seafloor.

Other geological evidence points towards an even younger date. An analysis of detrital zircons from the Jack Hills terrane in Western Australia revealed a chemical composition indicative of the presence of liquid water during their formation. Zircons contain radioactive components and are great for age dating. The formation age of the Jack Hills zircons dates to 4.4 billion years ago, implying a hydrosphere with liquid water existed a mere 100 million years after our planet's formation. However, it takes a tremendous amount of water to fill an ocean. Where did it come from?

Earth is not the only place in our solar system where liquid water may be found. Geomorphic studies of Mars indicate that water once flowed on the surface of the red planet, and Jupiter's moon Europa is known for its hidden ocean of liquid water beneath the planet's icy surface. Water is not unique to Earth; rather, it is an integral component of planet formation. Perhaps water existed in the primordial debris cluttering our solar system before the planets formed. But how might we know?

Science builds upon data, measurements, and factual evidence. However, on a larger and grander scale, science is about creative visions, which we call hypotheses or theories. The human mind searches for patterns in the world around us, and once a pattern is recognized, an explanation is sought. A hypothesis is an untested idea born in the human mind, a possible solution to questions of how, when, and why. The hypothesis seeks verification by observing evidence from the world around us. Once a reasonable amount of objective evidence supporting the hypothesis emerges, a speculative theory develops to explain why, when, or how. With enough testing to confirm them, theories become facts. But the process is flawed at best. Newtonian physics was the final answer until it wasn't and quantum mechanics threw a wrench in the works, sending physicists scurrying back to their labs to create new theories.

Our understanding of events occurring billions of years ago relies on theories developed from data collected today. However, reliable data is sometimes sparse, so the preference for one theory over another depends upon the weight of evidence supporting each. One theory suggests that Earth's water originated from asteroids, comets, and other space rocks involved in the planet's formation. Today, we can easily measure the ratio of deuterium to hydrogen (DH ratio) in Earth's oceans. That ratio may increase over time as hydrogen, the lighter of the two, leaks into space, but it should not decrease. Comparing the DH ratio of common seawater to the DH ratio of water in asteroids and comets is a way to test the theory. However, collecting water from objects in outer space is a challenging task.

Asteroids and Comets

Asteroids and comets are both familiar celestial objects, but they are structured differently. Asteroids contain a solid core consisting mainly of rocky material and metals, whereas comets are mostly ice, dust, and smaller amounts of rocky material. Water in the form of ice is a known component of comets. But the DH ratio of water in comets ranges from

1 to 3 times higher than Earth's water. Remember, DH ratios can grow over time but not shrink, so comets are not the primary source of Earth's water. However, asteroids are a better fit for the theory.

Asteroids are split into three categories:

C-type (chondrite): most common type of asteroid; made of silica rock and carbonaceous materials

S-type (stony): made of silicate materials and nickel-iron

M-type (Metallic): made of nickel-iron only

The opportunity to test the theory that water came to Earth via asteroids arose as a new wave of space exploration took hold. The ensuing physical sampling of tiny celestial bodies has helped us understand the origin of our planet.

Asteroids are small compared to Earth, making them hard to see and even harder to study. But we followed the old adage, "*If the mountain will not come to Muhammad, then Muhammad must go to the mountain.*" The asteroids would not come to us, at least not in pristine condition, so we sent remote spacecraft to them to sample these space rocks and return the samples to Earth.

JAXA (the Japan Aerospace Exploration Agency) was the first to collect particles from an asteroid in space when their Hayabusa mission in 2010 retrieved samples from the asteroid 25143 Itokawa. Later, in 2014, the Hayabusa2 mission was launched, and it eventually returned to Earth with samples of primordial material from the asteroid 162173 Ryugu. Water with DH ratios similar to ocean water was found in the Hayabusa2 samples from Ryugu. The asteroid is a C-type or chondrite, and these type of asteroids are known to have DH ratios similar to, but slightly higher than, Earth's water.

This DH ratio evidence generally supports a popular theory that our planet's water originally came from outer space as waves of asteroids inundated early Earth about 700 to 800 million years after the planet formed, during a period known as the Late Heavy Bombardment. However, theories are simply statements of probability and not necessarily definitive facts, so the Late Heavy Bombardment Theory doesn't necessarily jive with observations from the Jack Hills zircons that liquid water was present 4.4 billion years ago. Perhaps there was an earlier phase of bombardment we cannot now detect.

Further research explored another idea regarding why Earth's water is slightly lighter than that of asteroids. One of the benefits of scientific research is that its methods allow for unexpected results. So, when JAXA granted a group of researchers access to three rare pieces of the Itokawa asteroid from the original Hayabusa mission, the researchers hoped to investigate space weathering on the outer dust particles of the asteroid. Technology played a significant role in their research, leading to an unexpected discovery. The research team employed a novel analytical technique known as atom probe tomography. This process enables the analysis of individual atoms and molecules, determining their positions within a 3D matrix.

During their work, they unexpectedly found water. Specifically, they found a layer near the asteroid's surface that was rich in both hydroxide (OH) and water (H2O). In the true spirit of scientific research, they pondered the evidence for a bit.

The quandary they faced was that all previous research indicated this near-surface layer should be "*dry as a bone.*" The presence of water indicated the asteroid was receiving new hydrogen atoms from somewhere. The largest and most prolific source of hydrogen in the solar system is the sun. We also know the sun constantly produces coronal mass ejections, sending blasts of protons and electrons hurtling into space at a million miles per hour or faster. This process gives rise to solar winds.

Remember, a hydrogen atom is the simplest of atomic structures with one proton and one electron. We also know some of the protons riding a solar wind can scavenge an electron on their journey, creating hydrogen. The evidence for this phenomenon was provided by NASA's Imager for Magnetopause to Aurora Global Exploration spacecraft, which carried a Low Energy Neutral Atom (LENA) imager. Measurements confirmed that about one in every ten thousand protons in a solar wind snatches an electron on its journey to Earth, creating hydrogen.

This process of hydrogen creation would have allowed the new hydrogen molecules arriving with the solar wind to combine with the hydroxide on the asteroid's surface to create new light water. This excess light water possibly lowers the overall asteroid DH to match the DH ratio of Earth's water.

Several data points alone do not prove the theory, but they certainly strengthen it. What has been determined is that Earth's water has a DH ratio similar to water from asteroids, whereas the water from comets is much heavier than water in Earth's oceans today. The data support the theory that Earth's water came from asteroids.

So, in practical terms, water may have been present at the planet's inception, which means that life had the chance to get off to a quick start on a world rich in water and optimally located to keep from either frying in the sun's heat or freezing in the cold of deep space. The stage was set, and all that life needed to do was miraculously emerge. Perhaps our planet's generous supply of water is a gift from outer space. When we pollute our oceans, we should remember that we are defiling the precious gift that makes life possible.

Abiogenesis

In the beginning, there was the Void, formless and unfathomable.

The Void groaned and gave birth to the universe, and the universe spewed out life.

Water defines Earth. It provides a foundation for the thin shell of life on our planet's crust, where the biosphere thrives. Exactly how life emerged is a mystery. We can only peer backward through time to when life left its first fingerprints in the geological record. Life clearly started before then, but the specifics elude us.

The oldest uncontroversial evidence of life lies frozen in stone, recorded in rocks from Western Australia, a remnant of life from 3.4 billion years ago. These ancient remains were originally bacteria, specifically cyanobacteria, simple creatures that appear in the geological record as fossil stromatolites. If you aren't a stromatolite enthusiast, picture something like a large head of cabbage buried in sediments and transformed into rock. When you break open the rock 3.4 billion years later, you can recognize the rounded shape of a bioherm composed of thin layers stacked one upon another. Each layer represents a colony of single-cell organisms that grew and flourished in the shallow oceans of the Archean Era. When storms and waves stirred up the ocean-bottom sediments into a turbid sea, small grains of sand and silt were captured by the colony. When the cells died, the sand and silt remained, forming the substrate for the next layer of living cells. This process eventually created the layers we now see as a fossil stromatolite.

Biologists understand the basic building blocks of our biosphere, and we know that the appearance of Earth's first living organisms marked the start of a long evolutionary journey. However, this understanding does not answer the question of how. The exact origins of life's fundamental building blocks have vanished in the primordial history of our planet. Perhaps inception occurred on a deep ocean floor, in the extreme pressures and hellish temperatures surrounding a hydrothermal vent. Or it may have occurred in a warm puddle of slime as electric currents pulsed through the water from lightning strikes. Some even argue it fell to Earth from outer space.

Regardless of its origins, we know that deoxyribonucleic acid (DNA) provides the virile foundation of life shared by nearly all living organisms. Those few self-replicating organisms without DNA, like viruses and prions, use DNA's close companion RNA (ribonucleic acid) to accomplish the task.

The genius of DNA is its embodiment of the primal nature of existence, a quality that life cannot do without. Its mantra could be characterized as, "*If one of me is good, then two of me are even better.*" Without the ability to self-replicate, life could not thrive and evolve.

Water and DNA

The structure and nature of DNA provide us with some hints about the conditions surrounding its formation. The molecular double helix never occurs in isolation. It is always coated with water molecules, clinging to the DNA with the help of hydrogen bonds. Water is essential for the proper functioning of DNA and, hence, for the propagation of life. As I have previously stated, "*Without water, there is no life.*"

DNA also has temperature requirements. Above its melting temperature, the two strands of the double helix will separate. DNA melting, also known as DNA denaturation, is dependent on numerous factors and is influenced by the melting point. Denaturation for most DNA begins at about 52 degrees Celsius. The type of DNA also matters, and some DNA is more stable at higher temperatures. Various organisms live in temperature ranges that are outside the norm. Psychrophiles thrive at temperatures around 0 degrees Celsius, while thermophilic organisms survive at temperatures up to 100 degrees Celsius

Scientific literature contains an array of theories proposing mechanisms for DNA formation. A classic explanation portrays life, hence DNA, originating in a pool of water rich with organic compounds. The heat from the sun provides the necessary energy to mix things up and get the ball rolling.

Others point out that life would quickly go extinct in a pool with a limited volume of water since life processes would rapidly deplete the available chemical and nutrient base. What is needed, they insist, is a constant supply of the chemicals and compounds necessary to sustain the continuous growth of a population. Subsea hydrothermal and volcanic vents are one obvious source of this nutrient supply. This thinking leads to theories about life developing around deep-sea vents where heat and raw chemicals from deep in the Earth are readily available.

We know that today in the black, cold waters far below the ocean's surface, vibrant, diverse communities of living organisms develop around hydrothermal vents. The concentration of heat energy and chemical compounds, which are used in metabolic respiration, provides all the necessary conditions for life to thrive. Usable energy is the key that opens the door for life to take hold.

The Problem of Entropy

Entropy is a thermodynamic quantity representing the unavailability of a system's thermal energy for conversion into mechanical work. It is often defined as the degree of disorder or randomness in the system.

One of the problems facing all attempts at explaining the origins of life is the second law of thermodynamics: The total entropy of a system either increases or remains constant in any spontaneous process; it never decreases. Stated another way, it suggests that all systems will tend towards increasing randomness and decreasing order. This idea is counterintuitive to our lived experience. Clearly, a living creature represents a more complex, orderly system than a bucket filled with all the chemicals in a human body. Life appears to temporarily defy the second law of thermodynamics by increasing order and decreasing randomness.

Yet, almost everyone putting serious thought into the origins of life agrees that the presence of life must result from a spontaneous process that dissipates energy. This entropy constraint leads the theory of Thermodynamic Dissipation to explain the origin of DNA. A theory where DNA becomes nature's solution for dissipating radiant energy from the sun, thus facilitating entropy's demand for more randomness. DNA is the most efficient of all known molecules for absorbing intense ultraviolet light and transforming it into heat when liquid water is present. Both RNA and DNA efficiently absorb and dissipate ultraviolet light in the 230–290 nanometer wavelength range. This light spectrum range would have been part of the prebiotic, or pre-life, sunlight penetrating Earth's early atmosphere. Entropy demanded the most efficient pathway for the solar energy bombarding Earth's early atmosphere to dissipate and move toward the frigid cold of deep space. Nature's solution was DNA.

Perhaps the miracle of life is a reflection of a metastable universe. The biosphere of Earth is a steady-state, open, thermodynamic system (a dissipative system as defined by system control theory). Radiant energy, primarily from the sun, drives the biosphere. This energy provides the means to sustain life in a metastable system that will collapse when the sun's energy disappears. Under this line of thinking, life becomes the most efficient model for energy dissipation, given the particular temperature conditions of our planet.

Put more simply, DNA evolved as a means of increasing entropy. While this leads to locally decreasing entropy (increasing order) in the form of biological life, the net effect is an increase in the solar system's entropy through the dissipation of the sun's energy. The gist is that life spontaneously developed as part of a natural process. Given another water-rich solar system with a planet in the optimal temperature range for liquid water, we would expect it to happen again.

Something from Nothing

Abiogenesis refers to the theory that living organisms arose from inorganic substances billions of years ago. In Genesis 1, the first book of the bible, abiogenesis is described in poetic terms: "*God called the dry land earth, and the waters that were gathered together he called seas. And God saw that it was good. And God said, Let the Earth put forth vegetation, plants yielding seed, and fruit trees bearing fruit in which is their seed, each according to its kind, upon the Earth.*" (Life arises from a lifeless planet)

As the Hadean Eon ended about four billion years ago, DNA or its precursors became established in Earth's ancient oceans, and life was poised to take its first evolutionary step forward. Life needed to spontaneously arise from a lifeless ocean—abiogenesis. We discuss this process, but we don't fully comprehend it. Gaps in our knowledge obscure the details of the first life on our planet. However, before we engage in a discussion about first life, we need a framework for distinguishing when groups of molecules and chemical compounds transition from non-living matter to living organisms.

There is no set definition for what constitutes life, and instead, we must look to the basic qualities that differentiate flora and fauna from rocks. Most people broadly consider living organisms as complex, highly organized molecular assemblages with the ability to extract energy from their immediate environment and use it for growth and reproduction.

One of the fundamental concepts defining life is homeostasis, the ability of an organized system to maintain itself in a steady state despite changes in its external environment. A single cell has an internal environment that it must support. It must also have the ability to interact with conditions outside of its cell wall. As exterior conditions change, the cell monitors those changes and adjusts to maintain its internal environment. However, homeostasis can also be applied to certain non-living entities, such as organizations or industrial complexes. Thus, we need other qualities to pin down life.

Viruses Dead or Alive

Achieving homeostasis requires work, which involves acquiring and expending energy. We generally perceive living organisms as entities that engage in the acquisition, processing, and metabolizing of molecular compounds and other nutrients to generate energy. However, this requirement comes into question when we consider that a virus has no independent ability to metabolize and generate energy for reproduction. Is it alive, dead, or some sort of zombie life form?

The lowly virus appears to be another species trying to survive like the rest of life in Earth's biosphere. However, viruses occupy a fuzzy space on the edges of life. They are sometimes referred to as pseudo-living organisms because viruses don't have a cellular structure. They are simple organisms, but they still haunt humanity in the form of viral pandemics like COVID-19.

Almost all plant and animal life we see around us is cellular. Some organisms are single-cell creatures, but others, like humans, are multi-cellular, containing trillions of cells. Whether an organism is single-celled or multi-celled, it reproduces by cellular division. Animal cells have the metabolic machinery to create new cells. Viruses lack any metabolism and must reproduce by hijacking the cellular machinery of another organism.

Yet, a virus has aspects of cellular life. It possesses genetic material, reproduces, and evolves by natural selection. In these respects, it seems alive. The basic physical components of a virus are nucleic acid (DNA or RNA) wrapped in a protein shell. Viruses have a history on Earth as long as cellular life. They arose in the primordial mix of first life, and regardless of whether we consider them alive, they remain a vital part of the biosphere four billion years later.

The Mighty Cell

Given our latitude in defining Earth's earliest life, it is worth considering the division of simple life into two main categories: viral and cellular. Cellular life further breaks into three domains: Bacteria, archaea, and eukarya. Bacteria and archaea have an outer cell membrane, but their genetic material floats loose within the cell. The composition of their cell membranes differentiates the two. Eukarya also have an outer cell wall, but their genetic material is contained within a separate enclosed nucleus nestled within the outer cell wall.

We don't know which came first, viruses or cellular life. Some researchers posit that they both evolved from a common pseudo-cellular life form — cellular biology moved towards increasing complexity, and viral morphology regressed into a simpler form.

Speculation about Earth's earliest life forms necessitates accounting for the biological pathways required to access energy by either creating it internally or capturing it from another organism. Today, most organisms rely on free oxygen to produce energy. But, free oxygen would have been a scarce commodity in our planet's first oceans, so the earliest life forms must have consisted of anaerobic, chemosynthetic autotrophs.

Chemosynthetic refers to an organism's use of inorganic molecules as a metabolic energy source, and anaerobic indicates that metabolic activity occurs in the absence of oxygen. When an organism uses its metabolic energy to produce its carbon-based food, we call it an autotroph.

Life's first challenge was achieving homeostasis by adapting to conditions in Earth's oceans. Chemosynthetic bacteria were superbly suited to this task, and they were likely the first cellular life forms to unknowingly kick off a grand evolutionary party.

Fall of a Chemosynthetic Empire

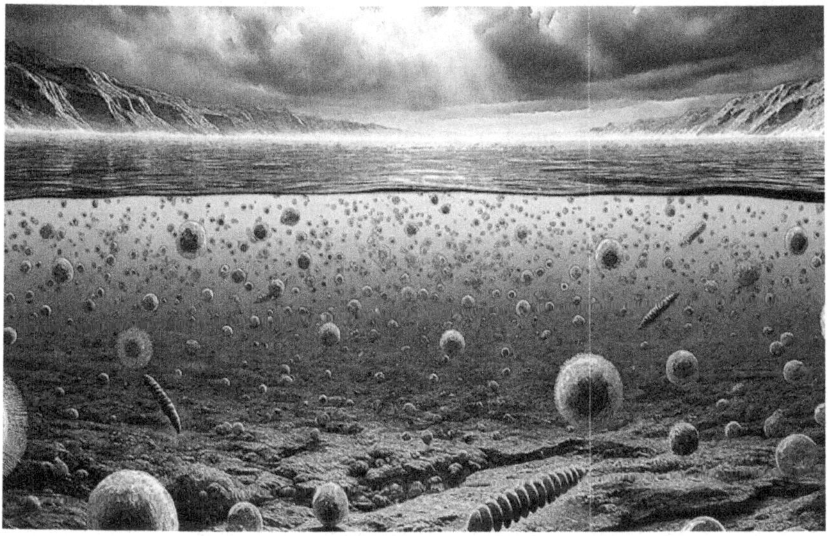

Evolution has no soul.

It creates new and beautiful forms of life and destroys that which cannot adapt.

It giveth and it taketh away.

Earth's first oceans were harsh and inhospitable compared to today's biosphere. There wasn't much for life to grab onto, only volatiles and gases absorbed in the primordial oceans during the initial degassing of Earth's interior: water, carbon dioxide, sulfur, nitrogen, chlorine, hydrogen, and other trace components. Notably, oxygen was not one of these components. Free oxygen (O_2) was not available for life to exploit. Early organisms needed to rely on other metabolic pathways, such as fermentation or anaerobic respiration, using nitrates, sulfates, or sulfur. However, metabolic activity using these molecules is much less efficient than cellular respiration using free oxygen. This inefficiency may have posed a limiting factor to the complexity of early life.

In addition to their metabolic inefficiencies, these first organisms were obligate anaerobes, which only survive in the absence of oxygen. Free oxygen was a fatal toxin for them. Nevertheless, these chemosynthetic bacteria were superbly suited to surviving under early-earth conditions and were probably the first form of cellular life.

Modern chemosynthetic bacteria thrive today in some of the most inhospitable environments on the Earth. They call hydrothermal vents at the bottom of the ocean home. The bone-crushing pressures and hellish temperatures around these vents make them happy. They also thrive deep in the polar ice, hidden in remote caves, and they even flourish in petroleum deposits. Chemosynthetic bacteria have been found in oil pipelines, where they eat from raw hydrocarbon flows and poop out sulfuric acid. They are tough little creatures and come with a set of skills that made them ideal for the harsh conditions on our young planet.

A good question would be, "Where is the evidence of this first ecosystem, this chemosynthetic empire?"

The answer is, "We don't have any evidence. The proposition is purely speculative."

Today, no evidence exists of their presence during the Hadean Eon and subsequent early Archean Eon—a vast stretch of time from 4.5 billion to 2.5 billion years ago. Likewise, their demise is obscured in the geological record. However, the case for chemosynthetic life forms being first comers to the biosphere is still strong. If they existed, their disappearance into the void of time constitutes our planet's first great mass extinction.

A possible catalyst for the disappearance of a chemosynthetic empire may be the lowly cyanobacteria.

Early Cyanobacteria

Scientists conducting research in Greenland in 2016 detected possible remnants of early life in rocks as old as 3.7 billion years. They identified signs of stromatolites in the rocks. Blue-green algae, or cyanobacteria, were believed to be responsible for creating the sediment layers, which became buried beneath thick piles of other sediments before eventually turning into rock. Importantly, the algae creating these rock layers were complex cellular organisms that may have been able to produce energy from sunlight via photosynthesis.

Allen Nutman and his research team visited Greenland in 2016, searching for evidence of ancient life. Before their expedition, the oldest discovered hints of early life came from the Strelley Pool Chert of western Australia — a rock formation containing 3.4 billion-year-old stromatolites.

Nutman and his colleagues traveled to the Isua Greenstone Belt, located in the Nuuk region on the southwest coast of Greenland. The town of Nuuk is situated on the coast, squeezed between the Labrador Sea and an inland ice cap. But between the town and the ice cap is a stretch of exposed rock called the Isukasia Terrane, where 3.7 billion-year-old rocks lie exposed at the Earth's surface.

Nutman and others believe the layered features they found in the Isua rocks are stromatolites. However, other early-life experts have posed valid questions about the findings, and we are awaiting more verification before casting a final opinion. Science is an arena where intelligent debate freely occurs—probably contributing to its absence in much government policymaking today. The investigations are still in progress, and the jury is out, but detectable, photosynthesizing, oxygen-producing life may have inhabited the early Earth's shallow seas 3.7 billion years ago. The occurrence would have been the beginning of the end for much of the chemosynthetic life at that time.

Free Energy from the Sun

In the dank waters of Earth's earliest oceans, single-celled chemosynthetic creatures carved out life's first niche on a newly formed planet, the third planet from the Sun. Perhaps they scraped by against all the odds, but natural processes related to entropy and thermodynamics probably necessitated their existence. They were alive by all definitions but woefully inefficient, pecking around in the dark and cold, looking for tasty inorganic compounds to consume and break apart. They stole energy from these compounds and used it to produce food for growth and reproduction.

It wasn't a cozy existence, and there is a reasonable chance these early bacteria spent their days crowded around volcanic vents and superheated hydrothermal flows, fighting over the best pickings. But life was life, and one did the best one could under the circumstances. Occasionally, these chemosynthetic floaters might have been fortunate enough to hitch a ride on an underwater current heading toward the surface. On a particularly good day, they might even bask in the glow of the young star we now call the Sun. Warm, radiant energy streaming down from above like mana from heaven would catch them in its ethereal glow. They could even absorb some of its heat if the radiation didn't kill them.

No neurons were firing in contemplative thought at the time, but if they were, these tiny primitive cells might have despaired at the unfairness of it all. Floating around in the cold and dark and endlessly struggling for food, when above them flowed a virtually unlimited supply of free energy. Radiant sunlight was there waiting for the cleverest of clever to tap into its power and produce food for free.

Today's autotrophs (organisms that produce their own food) rely on chloroplasts to take the Sun's free energy and convert it into sugars for food. These magical engines of energy transformation reside within individual cells. But like all aspects of life, chloroplasts didn't miraculously appear; they evolved into their current form. In a time long past, some single-celled, chemosynthetic prokaryotes discovered how to harness the Sun's energy by converting near-infrared light into sulfides or sulfate compounds.

The story most often told is that when the world was young, some archaea or bacteria managed to ingest other photosynthetic prokaryotes and steal their trick of converting sunlight into food. This turned out to be a repeating pattern in evolution: find something useful and quickly steal it for your personal benefit. My story is speculative, as we have no direct evidence of what happened; however, it seems to be a reasonable working hypothesis. The pigments from these early sunbathers were the initial predecessors to today's chlorophyll.

But oxygen was not yet part of the equation. Ancient bacteria produced food from the Sun, but their reactions used sulfides, not water. The revolution started when some of the bacteria figured out how to use water and sunlight to produce food by taking CO2 + H2O + sunlight and producing the sugars needed for life. However, all actions have unintentional consequences, and a byproduct of this particular version of photosynthesis was free oxygen (O2).

The first organisms to master this new technique were cyanobacteria, also known as blue-green algae. Despite being nicknamed '*blue-green algae*,' cyanobacteria are not algae at all. Cyanobacteria, and bacteria in general, are prokaryotes, whereas true algae are eukaryotes, and real algae developed long after the first cyanobacteria. However, the use of water in photosynthesis and the production of free oxygen as a byproduct forever altered the course of life's evolution. The lowly cyanobacteria started the blue-green revolution that eventually oxygenated our biosphere.

Since investigations of early life on Earth often involve more hypothetical thinking than hard evidence, there is considerable debate about when this blue-green revolution started. Conservative thinking puts the development of cyanobacteria at 2.7 billion years ago, a mere 300 million years before the great oxygenation event when the atmosphere gained significant amounts of free oxygen.

Other bolder lines of thought push the date back to between 3.4 to 3.7 billion years ago based on the presence of stromatolite fossil structures in Australia and Greenland. But there is a leap of faith in these interpretations. The fossil stromatolites are assumed to form from shallow-marine cyanobacteria bioherms, similar to the ones we see today in places like Shark Bay, Australia. But perhaps there were non-oxygen-producing precursors to cyanobacteria who liked hanging around together in bioherms.

The best estimate of when cyanobacteria became a significant part of the biosphere is rooted in the geological evidence of free oxygen. Free oxygen itself didn't fossilize, but its high chemical reactivity sparked various identifiable chemical changes in the geosphere and biosphere.

End of the Empire

Even though no direct evidence exists of an early Hadean/Archean world populated by chemosynthetic bacteria, a direct evolutionary jump from organic molecules to DNA-laden cyanobacteria seems unlikely. Such an evolutionary leap would be like Henry Ford setting up his factory and then having a modern-day Lamborghini as the first car off the assembly line. However, you don't go from Model T technology to a Lamborghini in one step.

These soft, single-celled, chemosynthetic organisms probably lived, died, and evolved for a billion years before the first photosynthesizing blue-green algae produced significant free oxygen. Unfortunately, these organisms had no defense against oxygen toxicity.

Their soft remains rained onto the ocean floor as they died and disintegrated, leaving no fossilized traces of life's earliest ecosystem. Perhaps up to a billion years of evolution disappeared as the biosphere oxygenated, wiping out most species in a hidden mass extinction event.

The true nature of the earliest organism on Earth remains a mystery. But abiogenesis worked its magic, and organic life appeared from an inorganic world. Early life wasn't much to look at—single-cell, chemosynthetic creatures and ancestral viruses floating in chemical-rich, primordial oceans. It wasn't pretty, but it was a start.

A Gateway to New Life

The history of humanity is recorded in books and lore.

The history of life on Earth is recorded in the rocks.

Inefficient metabolic pathways and poor cellular respiration were a fact of life in our planet's early oceans. Evolution did not progress beyond single-cell organisms during the first half of Earth's history. Life was waiting for a catalyst to open the door to a new future where complex life might thrive. That catalyst was free oxygen. Single oxygen atoms are notoriously unstable, and they quickly react with nitrogen, hydrogen, chlorine, and bromine compounds found in Earth's atmosphere and oceans, making the oxygen unavailable to organic creatures. Oxygen in a more stable form was required. Photosynthesis in ancient bacteria filled this need by producing free oxygen (two bound oxygen atoms) as a byproduct of converting the sun's energy to sugars for food.

Free oxygen in Earth's ancient oceans reacted with a variety of molecules to form new compounds, many of which were heavier than the water around them. Billions of years ago, as sufficient free oxygen became available in the oceans, these newly formed compounds accumulated in the seas and precipitated out of the briny waters as solids. Over time, deposits of these new materials formed in layers on the ocean floor and were buried, eventually becoming part of the rock record.

Today, geologists analyze these ancient rocks to determine when oxygen first appeared in the Earth's atmosphere. The appearance of free oxygen matters because it was the gateway to more efficient metabolism among Earth's population of single-celled inhabitants. We know that approximately 2.3 to 2.5 billion years ago, the oxygen content in rocks worldwide began to rise. The geological signature is so distinct that the period is known as the "*Great Oxidation Event.*" Some researchers have correlated the rise of cyanobacteria (also known as blue-green algae) with the timing of this oxidation event.

About a decade ago, research into South African rocks pushed the early oxygenation event back to approximately 3 billion years. One study looked at levels of molybdenum and iron in 2.95 billion-year-old sedimentary rocks. When free oxygen is available, it combines with manganese to form certain distinct molybdenum isotopes. The research on these South African rocks used the presence of oxygen-related molybdenum isotopes to conclude that these ancient sediments formed in an environment with significant free oxygen.

A second South African study investigated chromium isotopes in 3-billion-year-old rocks. The research results suggested that these chromium isotopes also formed in an environment where atmospheric oxygen was 100,000 times higher than could be explained by non-biological chemical reactions alone. These rocks whispered to us from over the eons, speaking of oxygen from photosynthesis in a primitive world whose origins are all but forgotten. These rocks provided us with information about our distant past and the first production of free oxygen.

The Pangola Supergroup

The rocks investigated in both South African studies were from the Pangola Supergroup. In geological terms, a group is a collection of geological formations that share certain lithological characteristics. A supergroup is just a collection of related geological groups. The Pangola Supergroup is one of the rare examples of rocks from the early Archean Eon that remain preserved in a relatively pristine state.

Plate tectonics puts a lot of wear and tear on rocks. Plate movement constantly pushes and pulls the continents, ensuring that most rocks are either raised up so weathering can grind them back into sand or buried deep underground where high pressures and temperatures squeeze and chemically alter them through metamorphism. The older the rocks, the more likely they have eroded away or changed into something new. However, occasionally, a group of rocks escapes this fate and remains relatively unchanged from its original condition.

The Pangola Supergroup is one of the lucky few whose sedimentary formations are still usable for studying a world long past. The Pangola Supergroup is situated on the southeastern edge of a larger geological feature known as the Kaapvaal Craton.

Kaapvaal Craton

The Kaapvaal Craton holds the secrets of early life on Earth. Entombed in its rocks are geological indications of single-cell prokaryotes and other simple organisms that gave life a firmer foothold on the planet. The craton underlies the northeastern third of South Africa and was formed during the Archean Eon between 3.6 and 2.7 billion years ago. But the oldest portions of the Kaapvaal Craton are exposed on its eastern edge in the Barberton Greenstone Belt.

Greenstone belts contain a variety of rock types. However, they are primarily composed of metamorphosed mafic and ultramafic volcanic rocks mixed with sedimentary formations. Both mafic and ultramafic are terms indicating that these igneous rocks have low silica content, typical of oceanic crust. Thus, greenstone belts are widely interpreted to represent the remains of ancient oceans, making sedimentary rock formations in greenstone belts very important in the search for early life.

Even without direct fossil evidence, chemical analysis of the Kaapvaal rocks suggests a partially oxygenated world existed approximately 3 billion years ago. Given the vagaries of rock preservation, it isn't unreasonable to think that oxygen-producing blue-green algae formed the Australian stromatolites, which clock in at 3.4 billion years. If this is true, then a billion years passed between the first free oxygen production and Earth's Great Oxygenation Event.

What was going on over this vast stretch of time, and why did it take so long for life to fully oxygenate our biosphere?

Science and Sulfur Isotopes

Science is not always big and bold. Sometimes, discoveries lie in the small and unseen. When volcanic gases react with the atmosphere, the reactions produce specific sulfur isotopes, which are eventually incorporated into rocks. Atmospheric composition plays a crucial role in determining the final form of these sulfur isotopes, and the presence of free oxygen leads to variations in the isotope's final chemical structure. So, in the geological record, an atmosphere with oxygen is distinguished from an anoxic (oxygen-free) atmosphere by analyzing sulfur isotopes.

Genming Luo and his associates at MIT used sulfur isotope technology to analyze rock cores from South Africa. Their investigations let them determine when the transition occurred from an anoxic atmosphere to a fully oxygenated one. These MIT scientists placed the timing of the Great Oxygenation Event at 2.33 billion years. They also believed that full oxygenation of the atmosphere took about ten million years. Wait a minute! If it only took ten million years for the transition, what happened to the oxygen produced over the previous billion years?

Concentrations of chemicals in the atmosphere, including oxygen, represent a balance between sources and sinks. An example of sources and sinks occurs when fossil fuels are burned, releasing carbon dioxide and becoming a CO2 source in the atmosphere. However, trees use CO2 to grow through photosynthesis; thus, trees are a sink. Blue-green algae was an active oxygen source in the early Archean, excreting free oxygen. If this oxygen wasn't collecting in the atmosphere, there must have been a chemical sink removing it.

Volcanic Gases and Earth's Mantle

A 2020 paper published in Nature by Shintaro Kadoya et al. proposes a solution to the mystery of the missing oxygen. The work by Kadoya indicates volcanic gases created a free oxygen sink, which was responsible for the missing oxygen. But that theory is just the tip of an iceberg whose roots extend deep below Earth's surface. The mantle beneath the Earth's crust is the source of magma for volcanic hotspots, such as Iceland and Hawaii, among others. Mantle plumes transport magma to Earth's surface, where the lava flows from volcanos, spewing a broad range of noxious gases into the atmosphere.

Kadoya's work focused on research suggesting that the composition of the mantle in the Archean Eon differed from today's and that ancient Earth's mantle contained less oxygen. A less-oxidized mantle produces higher volumes of gases that are prone to react with oxygen in the atmosphere. Gases, such as hydrogen, naturally combine with free oxygen, thereby removing it from the atmosphere. He suggests that oxygenating the mantle took almost a quarter of Earth's history—a time during which a fierce tug-of-war between biology and geology took place. For a billion years, gases from active volcanoes gobbled up all the oxygen the biosphere could produce. Only after the mantle was fully oxygenated did the oxygen-reactive gases stop flooding our atmosphere. This change allowed oxygen to accumulate, opening an evolutionary pathway for oxygen-breathing species to develop and thrive.

However, the mantle oxygenation scenario is not a fact, only a theory. Another line of thought views the formation of supercontinents as the key to understanding the Great Oxygenation Event. This theory builds on the biological needs of single-celled organisms and pinpoints nutrient supplies as the critical link for understanding why Earth's oxygenation took so long.

Mountain Building and Nutrients

One of the modern environmental problems we face is the ecologically destructive effects of algal blooms. Often, these algal blooms are initiated by excess pollution. The two primary culprits are agricultural runoff and leakage from waste treatment facilities in urban areas. Both sources deliver excess nutrients into our waterways. Under normal circumstances, algae populations are self-limiting because when they grow too large, the algae use all available nutrients. Once critical nutrients become scarce, the population declines.

But, an excess of nutrients allows an algae population to keep growing. This growth sets off a chain reaction where so many algae are dying that the bacteria trying to digest the dead cells consume all the available oxygen in the water, leading to a die-off of fish and other aquatic life.

We can flip this situation around and examine what happens when nutrients are always scarce. This scenario leads to perpetually small populations of algae, or in the case of the Archean Eon, blue-green algae. If nutrients were scarce through most of the Archean Eon, then there were never enough nutrients for a blue-green algae population explosion. In that case, oxygen production would have been perpetually low until the nutrient supply increased.

The collision of continental plates is one way to create a larger nutrient supply. These collisions create mountains, and as soon as mountains rise, erosion starts wearing them down. Weathering and erosion of the mountain rocks generate a supply of fresh nutrients for streams and rivers to carry to the ocean. This alternative theory proposes that the onset of the Great Oxidation Event was triggered by a rapid release of nutrients associated with increased tectonic activity. Subsequently, a steady supply of nutrients from continental weathering led to the growth of massive oxygen-producing algae populations, creating a surplus of free oxygen in the atmosphere. This theory is supported by numerous bits of research that point to the end of the Archean Eon as a transition period from a passive tectonic mode to the onset of sustained plate tectonics.

Rocks Tell the Story of Oxygen

One of the great mysteries of Earth's early development lies in the seemingly late oxygenation of our atmosphere and oceans. Theories, like those discussed above, must suffice because theories are all we have. However, the evidence suggests that between 2.4 and 2.3 billion years ago, our planet experienced a dramatic increase in the levels of free oxygen. Before then, free oxygen was sparse, and low concentrations of this key ingredient limited the ability of life to evolve into complex multi-cellular organisms.

Understanding the history of Earth's free oxygen is intimately tied to ancient geological formations called "*banded iron formations.*" These rocks display beautiful layering, alternating between black and red bands. The importance of this layering derives from how the red bands get their color. The red comes from the inclusion of tiny iron oxide particles in the original sedimentary matrix.

Iron oxide is exactly what the name implies: a compound formed when iron reacts with free oxygen. Think of it as ancient rust. Banded iron formations are major sources of iron ore for modern industry, but the last of these chemical sedimentary rocks formed over 800 million years ago.

The history of banded iron formations is also the history of our planet's journey from anaerobic to aerobic ecosystems. These formations trace the fluctuating levels of free oxygen in Earth's atmosphere and oceans.

Banded Iron

Banded iron formations are chemically derived sedimentary deposits. We often think of sedimentary rocks as sands, silts, and clays carried by rivers and streams into oceans and lakes. The sediments accumulate in layers and eventually harden to form solid rocks under the high temperatures and pressures of burial.

However, chemically derived deposits don't originate on land; they form through chemical reactions in the oceans. The small particles from these reactions slowly drift to the ocean bottom, forming stratigraphic layers.

Banded iron formations rely on the presence of iron and the availability of free oxygen to chemically react with it. The oxidized iron particles precipitate out of solution and drop to the ocean bottom. When the presence of free oxygen is sporadic, there are distinct periods where iron oxides form, interspersed with periods where no free oxygen is available, and other types of sediments are deposited on the ocean bottom. This process leads to the visual red and black banding for which these rocks are famously known.

So, banded iron formations only occur when oxygen levels are low enough for all free oxygen to be periodically depleted. Ancient oceans would have alternated between aerobic and anoxic conditions. This characteristic makes banded iron formations direct indicators of oceanic oxygen levels, providing us with a record of Earth's oxygenation.

The History of Earth's Oxygenation

The geologic record shows how Earth slowly became oxygenated. The oldest known banded iron formations are found in Greenland's Isua Greenstone Belt and are tentatively dated at 3.7 to 3.8 billion years. However, these types of deposits did not become widespread until the late Archean, between 2.8 to 2.5 billion years ago. The global disappearance of banded iron formations just after the end of the Archean Eon is critical evidence for the Great Oxidation Event. Once oxygen was constantly available in the oceans, the banded iron deposits could no longer form.

Interestingly, two periods of banded iron deposition occurred after the Great Oxidation Event, serving to remind us that Earth's environment was still in a state of flux. Unknown environmental conditions depleted the oceans of free oxygen approximately 1.8 billion years ago and again around 800 million years ago.

We can say with some degree of certainty that multi-cellular life did not make an appearance on the stage of life until after free oxygen was readily available.

The Cryogenian Big Freeze

The biosphere is a treasure Gaia nurtures from her breast.

We all drink the milk of life, but too few appreciate the gift.

The Magic of Oxygen

Without oxygen, most complex life would cease. The element oxygen traces its history back some 13.3 billion years to a period a mere 500 million years after the universe came into existence. Astronomers peering into the past analyzed light from a distant galaxy, and in this ancient light was the radiation signature emitted by glowing oxygen atoms. The light in question arrived on Earth 13.28 billion years after being emitted, making it the oldest record of oxygen in our universe.

Oxygen atoms form deep within the interior of stars, where the intense pressures and extreme temperatures of nuclear fusion shape them. When these stars explode, they spew their oxygen across vast reaches of interstellar space. This oxygen eventually collapses into gravity wells along with other cosmic debris, forming planets. The reality is that elemental oxygen has been present on Earth since its inception. However, when we discuss oxygen and life, we are not referring to elemental oxygen; instead, we are referring to molecular oxygen (O2).

A single oxygen atom has limited relevance for cellular metabolism, but when two of them are combined, the resulting molecule (O2) becomes the grand enabler of modern life. Anaerobic cellular metabolism utilizes inorganic molecules, such as ammonia, methane, or hydrogen sulfide, and combines them with elemental oxygen to produce energy. But the process is inefficient. Molecular oxygen (O2) provides a significant efficiency boost to cellular metabolic processes.

Using molecular oxygen, aerobic metabolism is eighteen times more efficient than anaerobic metabolism in producing ATP (Adenosine Triphosphate – the cell's energy currency). The introduction of free oxygen (O2) into Earth's atmosphere and oceans provided evolution with a tool for producing more complex lifeforms.

A Slow Process

The term oxygenated atmosphere only conveys the message that some appreciable amount of free oxygen was present; it does not specify what constitutes an "*appreciable amount.*" Oxygen levels at the end of the Great Oxidation Event were very low compared to modern-day levels. According to some researchers, oxygen did not reach its modern-day levels until relatively late in the evolutionary process, about 450 million years ago.

Given this late arrival of a fully oxygenated Earth, eukaryotes must have muddled around for a billion years, developing their survival skills before life as we know it arose. Even today, we know some eukaryotes can thrive in low-oxygen niches.

We read earlier that banded iron formations are geological indicators of alternating oxygenated and anoxic conditions in the oceans, telling a story of periodic regressions where free oxygen became scarce. Massive banded iron deposits appeared at the end of the Archean Eon and were last deposited on a significant scale approximately 800 million years ago.

The Proterozoic Eon followed the Archean, and this long interval is split into three eras. The Paleoproterozoic from 2.5 billion to 1.6 billion years ago, the Mesoproterozoic from 1.6 to 1 billion years ago, and the Neoproterozic, which ended 541 million years ago. So, the last period of widespread anoxia occurred during the Neoproterozic Era.

But the presence of these Neoproterozoic banded iron formations does not necessarily imply a world with no free oxygen. Clearly, portions of Earth's oceans turned anoxic, but this does not mean total anoxia prevailed throughout all of the planet's oceans. Environmental factors related to plate tectonics and volcanism may have regionally depleted the oxygen in some large bodies of water without affecting other areas.

Researchers recently identified banded iron formations in Western China, which date back to 527 million years before present day. These rocks imply anoxic conditions, at least locally, well after the Cambrian Explosion when all the branches of modern life established themselves. Oxygen was and still is extremely important to life as we know it, but the process of fully oxygenating our atmosphere was a slow, plodding affair.

Eukaryotes

Eukaryotic cells are the foundation for most complex life forms on our planet. The plants and larger animals surrounding us all require eukaryotic cells, but no settled timing exists for their exact appearance in the evolutionary timeline. Prior to the Great Oxidation Event, prokaryotes and archaea were the cellular building blocks for life. A generally accepted time range for the appearance of eukaryotes is between 2.3 and 1.6 billion years ago, potentially a billion years after Earth's oxygenation.

The introduction of significant free oxygen into Earth's biosphere was unquestionably a seminal event in life's evolutionary journey. However, the billion-year gap between the eukaryotes' first appearance and the emergence of widespread, complex, multicellular life remains an enigma.

The geological record remains foggy on this account, but several ideas deserve some speculative thought. Perhaps evolution is simply an exponential process, requiring a long build-up to critical mass. Homo sapiens emerged some 300,000 years ago and spent the next 295,000 years building up to the earliest human civilizations. Building momentum often takes time. However, that's not always how it works. We have the Cambrian Explosion, during which all modern biological lineages appeared in a geological blink of an eye. Afterward, the rate of core biological diversification mysteriously slowed in an un-exponential way.

But there are other possibilities. Perhaps free oxygen was not enough, and additional help was required. Efficient cellular metabolism provided a huge advantage for eukaryote development and evolution, but we know from modern-day observations that other factors like nutrient availability sometimes limit species development. Is it possible that nutrient scarcity put a damper on evolution's joy ride?

We don't know the answer. What we do know is that an oxygenated planet was necessary for the eukaryotes on which complex life depended. After the eukaryotes appeared, another billion years passed before evolution fully capitalized on these marvelous cell types.

An odd cold spell descended upon our planet during the middle of the Neoproterozoic. The period from 720 million to 635 million years ago is known as the Cryogenian, aka Snowball Earth. At the end of the thaw, soft-bodied multicellular life forms had gained a foothold in Earth's oceans. Something had changed.

Hot and Cold

Earth has run hot and cold for 4.5 billion years, but a warm planet is the norm. Earth has spent about 670 million years of its history in glacial periods, so cold spells account for only 15% of its existence. Earth's normal state is a hot, ice-free planet. Technically, we are currently experiencing a cold spell, but we are rapidly transitioning back to hot Earth conditions due to anthropogenic warming. We are now emerging from the last ice age, which was quite cold, with massive glaciers covering much of the northern hemisphere. However, 720 million years ago, our planet plunged into an unimaginable deep freeze during a period known as the Cryogenian.

Atmospheric carbon dioxide (CO2) is measured in parts per million (ppm), and currently, Earth averages about 420 ppm of CO2 in its atmosphere. I once had someone comment that CO2 certainly couldn't make much of a difference in temperature at such low concentrations. After all, even a change of 100 ppm is just a 0.01 percent alteration. They were arguing against the reality of climate change. However, small concentrations of CO_2 are deceptive because they have an oversized effect on atmospheric heat retention. Perhaps the best way to understand the problem is by pondering what the world would be like if our current 420 ppm level of CO_2 fell to zero.

Global temperatures today average about 15 degrees Celsius (59 Fahrenheit). If we removed all greenhouse gases from our atmosphere, with CO_2 being the primary warming component, average global temperatures would plummet to minus 18 degrees Celsius (-0.4 Fahrenheit). Freshwater freezes at zero degrees C, and seawater freezes at -2 degrees Celsius. Without greenhouse gases, Earth would become an ice world.

The Blue Planet

Today, Earth is a blue jewel floating in the dark vacuum of space. A stunningly beautiful planet we often see in pictures from satellites. We live on a planet with 71 percent of its surface covered by water. The oceans absorb most of the color from the light spectrum and reflect predominately blue wavelength radiation. However, a blue Earth would not be the case without greenhouse gases. Earth would look like a sparkling white diamond suspended against the deep, black background of the universe. An ice-encrusted planet, shining like a snowball in space. This is why the Cryogenian period is also known as "*Snowball Earth.*"

The Cryogenian period began 720 million years ago and lasted for 85 million years. The period occupied the middle of the Neoproterozoic Era and ended with the start of the Ediacaran Period. Geologists divide the Cryogenian into two phases. The early Sturtian glaciation event, spanning from 720 million to 660 million years ago, was the longest of the two. It was followed by the Marinoan glaciation, which ended 635 million years ago.

The debate about conditions on Earth during the Cryogenian alters between a planet frozen solid from the poles to the equator to a world where equatorial oceans existed as an icy slush. However, computer simulations indicate average equatorial temperatures of -23 Celsius or lower during the Cryogenian, supporting an ice-bound globe. Regardless of whether we discuss a frozen-solid Earth or a slushy one, such conditions were not static, and undoubtedly, the climate fluctuated over time.

The extreme climatic conditions during the Cryogenian Period would have stressed young Earth's biosphere. However, the expansive length of time over which Cryogenian glaciation occurred allowed evolution to cope with the adverse environmental conditions.

Consensus

There is no consensus on the causes of Cryogenian glaciation, but there is general agreement that fluctuations in solar radiation and Earth's carbon cycle are both critical factors. The onset of icy Cryogenian conditions coincides with the breakup of the Rodinia supercontinent. During the early Neoproterozoic Earth's continental plates were welded together into a single landmass that broke apart about 750 million years ago. Some theories hold that the breakup promoted excessive continental weathering. As rocks break down from the weathering process, significant CO2 is absorbed, thereby reducing atmospheric CO2. As Cryogenian CO2 levels dropped, the planet cooled.

The normal geological carbon cycle involves the injection of CO2 into the atmosphere from volcanic eruptions and basalt flows. This process is counterbalanced by the chemical weathering of silicate continental rocks, which removes CO2 from the air and converts it to dissolved bicarbonate. The carbon cycle involves feedback loops whereby increases in atmospheric CO2 drive global warming, and the warmer climate encourages increased chemical weathering, thus removing the extra CO2. As the CO2 disappears, the planet cools again. However, when this process becomes unbalanced, the planet can become excessively hot or cold depending on whether CO2 is being added or removed from the atmosphere.

We must also consider that rapid planetary cooling promotes more glaciation. The resulting ice and snow have a higher albedo than rock and open water and reflect more sunlight. Less solar radiation is then absorbed by the planet, further contributing to a global drop in temperature. The inference is that chemical weathering and reduced solar radiation conspired to cause Snowball Earth.

These theories are speculative and the exact causes of Cryogenian glaciation are not certain, but the geological record confirms that the planet plummeted into a deep freeze. When Earth thawed, life had mysteriously taken a leap forward, and the first simple multi-celled animals appeared in our oceans, sponges. Sponges are simple, stable creatures with life spans of up to 10,000 years and an ability to adapt to a vast range of environmental conditions. Evolution picked well.

The farther back in time we go, the greater our difficulty in identifying fossilized animal remains because there were no hard parts like bones or shells to preserve in the earliest fossil record. However, many scientists believe sponge-like creatures were the first animals on the zoological evolutionary tree. Yes, somewhere in the dim annals of history, your earliest ancestors were brainless sponge-like creatures hanging around together, thinking about absolutely nothing.

The evidence for this claim is based on molecular phylogeny, a technique using DNA molecules to determine the evolutionary history of genes and species. Extrapolations from phylogenomic analyses suggest that the first ancestral sponges emerged during the Cryogenian. Perhaps the young earth creationists will now be happier knowing that humans descended from a sponge, so we can remove ourselves from all this Scopes Trial controversy about mankind descending from apes.

These ancient sponge creatures lacked true hard parts, but they likely had tough protein fibers called spongin, which formed a three-dimensional network. Spongin is preserved in certain circumstances when it calcifies during the decay process after a sponge dies and is buried beneath sediments. This calcification leaves a fossil signature in the rock once the sediments solidify. Fossil signatures for sponges are well known in Cambrian limestones younger than 540 million years.

In 2015, researchers presented evidence of a well-preserved ancestral sponge fossil from a 600 million-year-old rock formation in South China. The fossil was exceptionally well preserved, allowing for the observation of cell types and cell structures in the remains. The fossilized animal was only about 3 millimeters in size, but it was unmistakable as an ancient sponge.

The Cryogenian Divide

The Cryogenian forms a dividing line in the evolution of our biosphere. Before then, single-cell life was the rule, but afterward, the planet thawed and we entered the Ediacaran Period, where complex, multi-celled organisms appeared and spread across the planet's oceans. Something happened in the cold, dark reaches below Earth's frozen oceans.

The Nature article '*Cryogenian magmatic activity and early life evolution*' (Jie Long, et al.) expresses this division well: "*Before the Cryogenian period, life was simple organisms, but afterward, evolutionary processes were characterized by a progressively accelerated stage, with life becoming more complex from 635 to 520 million years ago.*"

However, despite speculation and theories about a Cryogenian divide, we don't know whether life's step forward in the Cryogenian was simply coincidental or whether environmental conditions sparked a change. Maybe the groundwork was already in place for animal life when Sturtian glaciation kicked off.

Geologist and paleobiologist Elizabeth Turner (Laurentian University) believes she has uncovered a fossilized spongin network in 890-million-year-old rocks located in northern Canada. If she is correct, these fossils will be the earliest animal life yet discovered—animal life that preceded the Cryogenian.

The Cryogenian remains a significant divide, and evolution accelerated after this period, possibly due to changes in environmental conditions. The Cryogenian period followed the breakup of the Rodinia supercontinent. Tectonic plates separated and drifted apart, allowing magma to upwell from Earth's mantle, resulting in large-scale plate margin volcanism. A group of Cryogenian rocks in China named the Quinling Orogenic Belt records the geological history of a large magmatic-volcanic event.

Analysis of these volcanic rocks shows they are enriched in a variety of trace elements when compared to average felsic volcanics. Researchers studying the Quinling rocks proposed that increased access to these trace elements in the post-Rodinia oceans provided nutrients essential to the evolution of complex life.

It's an interesting proposition. If the planet's surface was encased in ice, then life might have retreated to areas deep in the ocean where magmatic or hydrothermal vents provided a continuous supply of heat and nutrients. Perhaps a combination of environmental stress and ample supplies of trace nutrients forced dramatic evolutionary changes as life struggled to cope with an ice-bound environment. Before the Cryogenian, photosynthetic algae dominated the biosphere, but their access to sunlight was dramatically curtailed as ice sheets spread from the poles to the equator. It was an environmental crisis and possibly a period of mass extinction.

One of the recurring patterns we observe throughout Earth's history is that rapid environmental changes stress life and lead to massive species extinctions. Life doesn't disappear, but it is severely choked back. Life returns once the crisis abates, but not in its prior form. After each mass extinction, we see the rise of new life forms and species. Perhaps the Cryogenian followed this pattern, and a world dominated by only algae was replaced by ocean floors covered with the new and wonderful organisms of the Ediacaran.

The Ediacaran

As the past recedes it becomes dreamlike.

We imagine more than we remember.

Darwin Didn't Know

When Darwin wrote his masterpiece "*On the Origin of Species*" in 1859, no fossils older than the Cambrian were recognized. It was not until the 1950s that definitive evidence of complex Precambrian life forms was confirmed. Another 30 years passed before Adolf Seilacher proposed that Ediacaran biota were not truly animals but rather a failed evolutionary branch he called '*Vendobionta*.'

Today, scientists are more inclined to view the Ediacaran as a period containing an eclectic group of organisms representing multiple evolutionary pathways — mother nature was rolling the dice to see what might turn up. Included within this grab bag of lifeforms were the early precursors of animal life as we know it today.

The Ediacaran Period occurred at the end of the Precambrian, spanning from approximately 635 to 541 million years ago. Earth had thawed from the Cryogenian deep freeze, and as the planet warmed, soft-bodied life took over the shallow oceans. The Ediacaran was a critical period when life mapped out new pathways like guts, limbs, and complex behaviors such as burrowing.

Our delay in recognizing 94 million years of evolution demonstrates that investigating Earth's ancient past is a tenuous affair. Uncovering evidence of life in the geological record relies on a combination of unusual preservation circumstances and the random luck of fossil deposits being at Earth's surface in the present day. One of the problems with preserving fossilized materials is the vicious competition for food within all ecosystems. Life takes dinner wherever it can be found and always devours it as completely as possible. All that is edible will be eaten as predators consume their prey and scavengers collect what is left. Bacteria then feast on any remaining organic debris. The biosphere always tries to recycle life into new life. This fact makes the chance of dead animal remains being fossilized very small indeed.

Ediacaran Biota

The biota of this period are named for the Ediacara Hills of South Australia. Organisms of the Ediacaran Period represent the first significant influx of metazoans into Earth's biosphere (metazoans are animals made up of more than one type of cell). The earliest metazoans probably made their appearance in the Cryogenian. Still, the 94 million years following the big thaw were when evolution honed these creatures' survival skills, allowing them to dominate their respective ecological niches.

Ediacaran animals primarily inhabited warm, shallow seas. They enjoyed the benefits of sunlight and the exchange of oxygen between the atmosphere and the ocean. Atmospheric and oceanic oxygen formed key components of the Ediacaran evolutionary saga. These early metazoans required free oxygen for cellular respiration, and oxygen may also have been a controlling factor in their preservation. Even though these organisms lacked any true hard parts, some of their cellular structures included collagen, making their tissue tougher, more resilient, and more prone to fossilization and preservation.

Interestingly, animals can't produce collagen when atmospheric oxygen levels are less than three percent of the modern levels. This fact leads to speculation that Earth passed this magical '*three percent mark*' during the Ediacaran Period, allowing evolution to incorporate collagen proteins into the cellular mix. Prior to reaching this mark, all fossil evidence consisted of trace fossils like burrows or trails, which let us know something was living but left us with no evidence of the kind of creatures that created the trace fossils. However, after reaching the three percent threshold, imprints of the animals were preserved in the fossil record.

An Early Peek at Animal Life

A 2018 article in Science magazine focused on a 558 million-year-old organism called Dickinsonia, a broad, flat, leaf-shaped creature. We are not talking about a microscopic critter. These organisms grew up to four feet in length. Dickinsonia fossils have been known for over 70 years, but until recently, debate raged about whether it was a large amoeba, an animal, or even a plant. Simply looking at the visible remains didn't make us any wiser.

However, the science of studying Precambrian fossils has evolved to include more than simple visual analysis for determining what might be an animal versus something else. Geobiologists have entered the academic fray and developed techniques to isolate fat molecules from Ediacaran fossils. This line of investigation identifies chemical products from the breakdown of cholesterol — a product common in animal cell membranes.

Another group utilized 3D laser scanning technology to identify tiny, fossilized Ediacaran creatures (Ikaria wariootia) with left-right symmetry and identifiable front and back ends. As much as we might not like to be reminded, a mouth, guts, and an anus are the basic defining characteristics of multicellular animal life, be it a worm or a human being.

Researchers studying Dickinsonia fossil remains isolated lipid biomarkers that were unambiguously indicative of animal life. This discovery makes Dickensonia one of the earliest identifiable Ediacaran animals in the geological record, but this doesn't mean it was the earliest animal, only the earliest recognizable fossil.

Mass Extinctions

The popular position on mass extinctions holds that we are currently in the midst of an Anthropocene extinction event, Earth's sixth mass extinction. Most literature places the first mass extinction in the Ordovician about 445 million years ago. However, this accounting appears to contradict paleontological evidence from the Late Ediacaran. Have we miscounted?

Mass extinctions are defined by the rapid disappearance of a majority of the biosphere's species. These catastrophic die-offs are recorded in the fossil record and form boundaries between distinct geological periods. One of the most definitive boundaries for early geologists was the transition from the Precambrian to the Cambrian, also known as the dividing line between the Proterozoic and the Paleozoic. As previously mentioned, the fossil record in Darwin's time began 541 million years ago, at the start of the Cambrian. For all appearances, we leaped from four billion years of no complex life to an abundance of life in the blink of an eye.

We now know that the beginning of the Cambrian period followed a long history of animal life during the Ediacaran Period. Mass extinctions are dramatically negative events when viewed purely from the perspective of the lost species. But mass extinction events don't purport to be total extinctions; life has always found its way to the other side. Throughout geological history, the period following a mass extinction corresponds to the rise of new species that fill the ecological void created by a massive die-off.

The Cambrian brought us an explosion of new life after a massive die-off: out with the old and in with the new. In this case, the old consisted of Ediacaran species thriving in the shallow oceans of our planet. During a brief period in the early Cambrian, almost all of today's animal phyla came into existence. Simultaneously, the world of Ediacaran biota slipped into the void of extinction, never to be seen again.

The global distribution of fossilized Ediacaran animals, along with their diversity, speaks to a biosphere rich in soft-bodied life. Preservation of some Ediacaran fossil assemblages is excellent, but relating them to modern phyla remains difficult. Still, we know life was thriving in Earth's shallow oceans for almost 40 million years before the Cambrian explosion.

Ediacaran evolution produced critical changes to animal life and introduced skills and characteristics like mobility, sexual reproduction, digestive systems, and more. The basic components of complex organisms developed in Ediacaran animals and prepared the biosphere for its jump into the Cambrian.

The Precambrian-Cambrian boundary marks the transition from soft-bodied life forms to complex animals with hard body parts. This change was rapid, and for the most part, fossils of Ediacaran animals are not found in Cambrian strata. Ediacaran ecosystems fell extinct and were replaced by the hard-shelled predators of the Cambrian. It sounds like a mass extinction to me.

A common theory suggests that the new Cambrian creatures were predators characterized by mobility, protective shells, and voracious appetites. They moved quickly compared to their sluggish predecessors and were able to win the daily battle between who eats and who gets eaten. Creatures on the Ediacaran seafloors stood no chance and were devoured into extinction. It's not an unreasonable theory, but perhaps more was going on.

The End of an Era

Discerning the cause, or various causes, of the end Ediacaran extinction is a puzzle that may never be completely solved, but not for lack of trying. Researchers in 2018 proposed the Ediacaran ended when Earth's oceans went completely anoxic. They analyzed uranium isotope variations in limestone samples from three well-known Ediacaran locales in China. The samples covered the last 10 million years of the Ediacaran Period, and once the isotope data was integrated with the fossil data, an episode of extreme anoxia was identified. The anoxia coincided with the disappearance of the Ediacaran biota. However, this data is specific to one locale and may not accurately represent global conditions.

Another slightly more subtle take on the idea of being devoured into extinction is biotic replacement. The earliest of the hard-shelled Cambrian creatures were probably burrowers and grazers. Their activities would have severely disrupted the Ediacaran lifestyle, interfering with food chains and destroying the existing seabed ecosystems. The theory is based on observing progressively lower species diversity during the last several million years of the Ediacaran, using fossil assemblages from a locale in southern Namibia. This scenario suggests a progressive extinction rather than a catastrophic one.

Ultimately, however, the weight of evidence points to an Ediacaran mass extinction caused by species competition in some form. It forms a partial analog for the ongoing Anthropocene extinction in this respect. A single species currently dominates Earth's ecosystems, Homo sapiens. Supporting a world population of over eight billion humans stresses the environment when the demand for natural resources soars and hungry mouths need feeding. Mass extinctions don't require catastrophic asteroid strikes or massive outpourings of magma from the mantle; they just need biotic replacement—by human activity in the case of the Anthropocene extinction.

The Cambrian, an Explosion of New Life

Evolution does not repeat itself.

It is like time, always moving in one direction: forward.

Thus far, we have traversed our way through the first four billion years of life on our planet as we explore our Vanished Origins. The farther back in time we travel, the less factual evidence we find, meaning we fill in the gaps with speculation and theories. However, the arrival of the Cambrian added more certainty to the geological record.

The start of the Cambrian Period, 541 million years ago, marked a pivotal point when Mother Nature stopped dawdling around. Evolution struggled for billions of years to produce single-celled bacteria and eukaryotes. Life was on the planet, but it was an inauspicious start to the complex biosphere we enjoy today. Finally, in the Ediacaran, complex multi-celled creatures made their way onto the stage of life. However, they were merely the warm-up act, and evolution shifted to hyperdrive with the arrival of the Cambrian.

Today, we recognize 36 animal phyla, but nine major phyla contain most of the extant species in our biosphere—Mollusca, Porifera, Cnidaria, Platyhelminthes, Nematoda, Annelida, Arthropoda, Echinodermata, and Chordata. All these major phyla appeared in the fossil record during the early Cambrian. So, after four billion years of struggling, the key branches of the animal kingdom evolved during a 15-million-year period.

If we reduce Earth's history to a single day, then all major branches of life appeared during a five-minute interval at about 9:30 in the evening. It's an extraordinary observation given our tendency to view evolution as a slow march over time, with life steadily moving toward increasing complexity.

The reality is that life exploded onto the scene during an incredibly short period of hyper-evolution. Sponges of the Porifera phylum appear to be the single member of our modern major phyla with a definable history before the Cambrian explosion. Evolution is not a steady flow of rising excitement. Instead, it is long periods of boredom punctuated by brief episodes of frantic change.

A Focus on Success

A focus on success defines evolution. Organisms best suited for the available ecological niches will inevitably thrive and come to dominate their ecosystem. We only know of them today because of their success. Failed attempts at species modification never make it into the geological record. Successful species populations grow, and the larger the population, the higher its chance of preservation in the fossil record.

To understand the characteristics of winners in early Cambrian evolution, we must note the rise of trilobites. For 55 million years, these creatures dominated the Cambrian biosphere, making them one of biology's great success stories. Their reign continued for another 230 million years after the end of the Cambrian, until their demise during the end of the Permian mass extinction—the Great Dying. Trilobites had hard exoskeletons for protection and limbs for motion and predation. The closest modern-day morphological analog is a horseshoe crab.

Their wild success was due to evolutionary innovations and species diversity. The term trilobite refers to a genus of animals with over 20,000 different species. They ranged in size from under a millimeter to over several feet in length and adapted to a vast range of habitats.

It is said that in the land of the blind, the one-eyed man is king. Trilobites were early innovators in the field of vision, with some species developing enlarged eyes. Imagine two creatures searching for the same dinner, but one can see it, and the other can't. You get the picture.

When the available ecological niche required swimming, fast bullet-shaped swimmers appeared. But when burrowing was the ticket to survival, trilobite species evolved to live in the sea-bottom mud. Some lost their eyes because they spent all their time underground.

Trilobites possessed a key ingredient of success: evolutionary adaptability.

Life in the Cambrian

The Cambrian was a great time to be alive if you were a trilobite. Living on land was not an option; life remained submerged beneath the ocean surface. Shallow marine environments were the hip place to be. There, you found sunlight and shallow water where oxygen from the atmosphere constantly seeped into the surface waters. These two combined benefits translated into '*plenty of food.*'

Earlier, before the Ediacaran period, the Rodinia supercontinent broke apart, and by the mid-Cambrian, two large landmasses had formed. The Gondwana supercontinent covered a vast area near the South Pole, while Laurentia was positioned closer to the equator. The shallow Cambrian seas along the continental shelves existed in the coldest and warmest areas on the planet. Adaptability was a must-have for species survival.

However, cold is a relative term. Global temperatures were rising during the Cambrian period. Previously, during the Neoproterozoic, Earth fell into a deep freeze during the Cryogenian period and was still recovering during the Ediacaran. Although the planet warmed during the Ediacaran, average global temperatures at the beginning of the

Cambrian were likely slightly lower than they are today. However, the Cambrian world continued to warm, and the average global temperature throughout the Cambrian was 22°C, as opposed to our modern-day average of about 15°C.

For context, Cambrian temperatures rose an average of 0.5 degrees Celsius per million years. In the Anthropocene, we have managed to achieve a rise of approximately 1 degree Celsius in Earth's average temperature in just 150 years. Unfortunately, our modern world is on a path toward a Cambrian-like condition. We are finding out the hard way that awareness of climate change is not enough. Action to mitigate global warming needs to follow our newfound awareness.

Why the Early Cambrian?

The fossil record is clear about a period of hyper-evolution in the early Cambrian. What is less clear is the cause of this anomalous event. In a universe of cause and effect, it's reasonable to suspect the Cambrian explosion relates to significant changes in environmental conditions. It is also likely that the early Cambrian evolutionary frenzy was stoked in part by increases in the availability of critical resources. So, the two prime suspects for change are oxygen and food/nutrients. Oxygen levels play a crucial role in the survival of individual organisms, and nutrients are essential for maintaining healthy ecosystems and large species populations.

The Cambrian period commenced approximately 2 billion years after the Great Oxidation Event. We also know the Ediacaran biota required oxygen, so an oxygenated atmosphere was present long before the Cambrian began. However, the degree of oxygenation at any particular time is difficult to determine.

The simplest atmospheric oxygenation models propose two major phases associated with rising atmospheric oxygen levels. The first was the Great Oxidation Event, where oxygen levels rose from negligible to measurable. The second step was a Neoproterozoic phase, during which oxygen levels rose sufficiently to support Ediacaran animal life. But this two-step model doesn't explain why the explosion of life didn't occur earlier in the Ediacaran.

Cambrian Oxygen

We can't look at the geological record for fossilized oxygen. Hence, studies of ancient environments use the presence or absence of oxygen-dependent compounds to decipher the mystery of oxygen levels. This practice requires rock samples spanning a continuous interval of time so the data provide a chronological sequence of measurements.

One such set of fossil-rich limestone strata in Siberia provides a continuous record of the early Cambrian. Researchers analyzed an extensive set of samples from these rocks, examining carbon and sulfur isotopes. Changes in these isotope levels indicated oxygen variations in the ancient Cambrian atmosphere.

They didn't detect a steady increase or decrease in oxygen levels. Instead, they uncovered a history of wild variations, creating a record of boom and bust oxygen fluctuations. The estimated average oxygen level during the early Cambrian was 8.4 percent by atmospheric volume. This level seems low compared to today, where atmospheric oxygen is about 21 percent. However, averages don't tell the whole story. The data from Siberia indicate that oxygen levels fluctuated by about 50 percent, meaning atmospheric oxygen varied between 4 and 12 percent during the early Cambrian period.

The Siberian research identified five oxygen spikes, with each peak corresponding to an increase in biodiversity. The oxygen lows correlated to high species extinction rates. These data suggest a coupling between higher oxygen levels and spurts accelerated evolution. However, we don't know exactly why more oxygen caused biodiversity to spike. One line of thought is that extra oxygen expanded the geographical range of ecosystems, giving life more opportunities to evolve. Another explanation correlates higher oxygen levels with increasingly complex predator development, creating an evolutionary arms race where the '*evolve or perish*' axiom necessitated rapid evolutionary change.

Nutrients

Another approach to understanding the Cambrian explosion focuses on food webs. The breakup of the Rodinia supercontinent in the Neoproterozoic and the subsequent formation of Gondwana and

Laurentia likely resulted in the creation of extensive coastlines and, consequently, vast shallow seas. Meanwhile, the tectonic collisions that formed these supercontinents led to uplift and mountain building. This geomorphic arrangement, along with a warmer climate and higher rates of weathering, led to massive continental erosion, and the eroded sediments were carried into the shallow Cambrian seas.

Perhaps this influx of nutrients stimulated plankton growth at the base of the food chain, creating a biological chain reaction as massive amounts of organic carbon entered the food web. The extra food provided energy for the scavengers and predators to flourish.

If we combine the oxygen story with a greatly expanded food web, we might have a plausible explanation for how Mother Nature rolled the dice in the early Cambrian and came up with a winner. Expanding ecosystems rich with food provided a lucrative environment for life to thrive and evolve.

An added bonus was that each dip in oxygen created an extinction event, and we know that the rise of new species and life forms often follows mass extinctions. The early Cambrian experienced five oxygen-related extinctions, or conversely, five opportunities for evolution to reset and develop something new.

The Cambrian explosion was a unique turning point in life's 4.5 billion-year journey. We still have more questions than answers about what happened during this brief 15 million-year period at the beginning of the Cambrian. We also don't fully understand why the original nine major phyla have remained and evolution has not replaced them with something even better. Maybe sticking with what works is a fundamental principle of evolution.

The Ordovician: Life Shifts Landward

Oceans are the womb of life.

It took four billion years before evolution broke free from the ocean to explore the brave new world of dry land.

The first insights into the Ordovician Period are credited to Charles Lapworth, who, in 1879, delineated a distinct set of fossilized fauna that differed from the preceding Cambrian Period and appeared separate from the younger Silurian fossils. Prior to that recognition, geologists were prone to follow their fundamentally argumentative nature and bicker about where the Cambrian ended and the Silurian began.

However, the Ordovician was not formally recognized as a distinct period until 1960, when the International Geological Congress adopted it as the second period of the Paleozoic Era, spanning approximately 45 million years, from 488.3 million years ago to 443.7 million years ago. During this period, much of the world's land was located in Gondwana, the southern supercontinent.

The end of the Cambrian party was not without drama, and life collapsed in a mass extinction event. College parties are famous for ending when the vital resource of alcohol becomes scarce, having been consumed voraciously for too long. The Cambrian also ended with the scarcity of a vital resource as oxygen levels collapsed, wiping out many species. Some researchers also believe that persistent anoxia in the late Cambrian oceans caused rising levels of hydrogen sulfide, a substance toxic to most marine life.

Despite the adverse environmental conditions at the end of the Cambrian Period, life clung on, maintaining a foothold on our planet. Evolution worked its magic, and those species not wiped out were joined by a diverse group of new marine invertebrate species, including graptolites, brachiopods, conodonts, cephalopods, corals, crinoids, and gastropods. Additionally, our old friends, the trilobites, regrouped and continued to flourish. Life thrived in the shallow Ordovician seas, and during this period, the first plants invaded the land.

Wide Shallow Seas

As in the Cambrian Period, shallow oceans remained the primary focus for Ordovician life. A combination of sunlight and oxygen was needed to keep evolution moving forward, and shallow, widespread oceans accommodated this need.

Temperatures steadily rose throughout the Cambrian and into the Ordovician. By the early to mid-Ordovician, Earth was a steamy, moist, hot planet with global atmospheric temperatures averaging about 49 Celsius (120 degrees F) and ocean temperatures reaching 43 Celsius at the equator. Remember that today we worry about an average global temperature of 15 Celsius. It was only towards the end of the Ordovician Period, about 443 million years ago, when Earth finally began to cool.

We don't need to engage in deep thinking to envision the Ordovician as a water world. There were no massive ice sheets, so all the planet's water was in Earth's oceans, filling them to the brim. Not only was there more water in the oceans than there is today, but the water also occupied more space.

In a 2019 report, NOAA tracked a global sea-level rise of 9

centimeters over a 25-year period. Thermal expansion due to increasing ocean temperatures accounted for 40% of the rise, and the remaining 60% was new meltwater from ice caps. Other research indicates seas may rise as much as 2.3 meters for each 1-degree Celsius increase in average ocean temperature. The math is staggering. The lack of ice sheets placed Ordovician seas 70 meters higher than today, and the thermal expansion added an additional 60 meters. Ordovician seas may have been over 400 feet higher than they are today.

Large portions of the proto-North American continent were underwater, but these conditions provided opportunities for evolution to progress. The vast areas of shallow marine habitat provided newly evolved species with room to expand and dominate their ecological niches.

Drifting Continents

The Earth was also undergoing geological reorganization during the Ordovician Period. Tectonic plates were on the move. The Gondwana supercontinent edged towards the South Pole, and Laurentia stabilized near the equator. Two other large landmasses, Baltica and Avalonia, slowly converged on Laurentia. They eventually collided in an event known as the Caledonian orogeny—a mountain-building event spanning the Ordovician and early Silurian.

The movement of Gondwana towards the South Pole was one of the triggers that eventually led to the demise of the Ordovician Period. Ironically, one of the hottest periods in Earth's history ended in the cold. A South Pole covered by land provided the substrate for glaciation and the formation of massive ice sheets.

The success of life during most of the Ordovician Period was driven by the availability of vast flooded continents, which formed shallow marine habitats for a dizzying array of new and successful species. They thrived in warm oceans under the light of an ancient sun. Ice was the enemy of these successful ecosystems. Ice cooled the planet and sucked valuable water from the seas, locking it away on land.

The assault was a twofold attack on the biosphere's critical habitats, wide shallow oceans. As water progressively accumulated in glacial ice sheets, sea levels dropped, and the atmosphere cooled. The cooling atmosphere sucked heat from the oceans, and simple physics worked against life as the cooler water occupied less space and further hastened falling sea levels.

As water drained off the drowned continents and created dry land again, the habitats for Ordovician life slowly disappeared. With shrinking resources, species competition increased, creating winners and losers. Some species went extinct so others could survive.

But another unsuspecting player entered into the fray, helping to usher in the well-known Ordovician-Silurian mass extinction—plants invaded dry land, changing the dynamics of chemical weathering and exacerbating global climate change.

Non-Vascular Plants

Life crept out of the oceans and onto dry land approximately 470 million years ago during the Ordovician Period. If you're imagining dappled sunlight dancing across the floor of a lush, primordial jungle teeming with tall, exotic trees and shaded by a thick canopy of leaves, you are wrong. Instead, picture a Martian-like landscape covered by rocks, boulders, and fine dust. But when you inspect more closely, small insignificant-looking plants appear, growing on the surfaces of the rocks and boulders strewn across the landscape. The ancient ancestors of moss and liverworts covered the Ordovician landscape. These first Ordovician plants were simple non-vascular organisms, but they had a major impact on the world.

Mosses, hornworts, and liverworts are all examples of non-vascular plants. They grow close to the ground with shallow root systems that could never support larger plants. Since they are non-vascular, their size is also limited by their inefficient internal transport system for water, gases, and nutrients. Vascular plants can pump water and nutrients from their roots deep underground to leafy canopies hundreds of feet in the air. Non-vascular plants are limited to living along the ground, close to their shallow roots. Still, the first plants on land were pioneers, and they had a fresh new world to exploit.

These early continental settlers reproduced by spores, not seeds. We know that seeds offer numerous advantages for plant reproduction, as their protective shell encloses nutrients and an embryo. The humble seed is a self-sufficient unit ready to produce new life and continue the species. Seeds are multicellular, but a spore is a single cell surrounded by a tough membrane. Spores are created asexually through a process called sporogenesis. They are not as durable and hardy as seeds, but each spore can grow into a new plant under the right conditions.

Life on a Barren Rock

Despite the sophistication of seeds, reproduction by spores had its advantages for these first Ordovician plants. Each spore is small, and many spores are easily transported by the wind. Seeds often rely on animals for transport to new locations, but there were no land animals in the Ordovician. Reproduction by spores provided a way for these new species to propagate themselves far and wide, filling Earth's ecological niches wherever the wind blew. Spores allowed these new organisms to conquer dry land.

One of the lessons learned by humans in the Anthropocene is the interconnected nature of our biosphere. Through a process popularly known as the '*butterfly effect,*' seemingly small changes can trigger a series of events that culminate in significant environmental alterations. The idea applies to deterministic non-linear systems.

Some complex dynamic systems exhibit unpredictable behaviors such that small variances in the initial conditions can have profound and widely divergent effects on the system's outcomes. Because of the sensitivity of these systems, outcomes are unpredictable. This idea became the basis for a branch of mathematics known as chaos theory, which has been applied in countless scenarios since its introduction.

Ordovician plants thrived on both solid rock and rocky soil. Although their roots were quite shallow, these first simple plants made a significant difference. In every location they occupied, they secreted a variety of organic acids into the soil and rocks below. This process was new. The norm on Earth's continents for four billion years was a land barren of life. However, during the Ordovician, this norm came under attack, and just like the butterfly effect, these small plants with tiny roots had a profound impact on Earth's climate.

One Thing Leads to Another—Chemical Weathering

These first Ordovician Period plants were like foreign invaders taking over the surface of Earth's continents. These plants figured out how to live outside of the oceans, bringing new life processes with them. By-products of their way of life altered the planet's surface. Acidic secretions from these simple plants enhanced the natural weathering process by accelerating the chemical breakdown of rock surfaces.

A research team at the University of Exeter, UK, grew moss on granite for 130 days and then measured the weathering effects. Observations from the mossy granite, when compared to a control granite surface with no plants, showed significantly more chemical weathering on the mossy rock. Non-vascular plants enhance weathering.

The roots of these ancient plants were shallow, and at first glance, small amounts of secreted organic acids may seem insignificant. But the world held unlimited potential for these plants, and with no predators to stop them, they spread across the planet, building our first dry-land ecosystems. The effects from individual plants were small, but eventually, their collective effect could not be ignored.

Faster weathering of the Earth's rock surfaces paved the way for another chemical reaction. As rock weathers and is chemically altered, it reacts with carbon dioxide (CO2) and removes it from the atmosphere. The moss and liverworts accelerated the weathering process, gradually helping to lower the CO2 content of the atmosphere.

Non-condensable greenhouse gases, like CO2, play a critical role in maintaining Earth's surface temperature in a range that supports life as we know it. Without these non-condensable gases, the Earth's average temperature today would be about -18 degrees Celsius. So, when weathering removed enough CO2, the first plants helped Earth slip into another glacial age, the Hirnantian ice age.

Silurian Magic

Evolution thrives on environmental adversity and species mortality.

The old must perish to make room for the new. A lesson worth thinking about.

A Lesson Worth Learning

The Ordovician Period was a great time for life on Earth until it wasn't. The biosphere was reminded of a biblical adage as it applies to mother nature: "*She giveth and she taketh away.*" The official end of the Ordovician clocks in at 443.8 million years ago, but the catastrophe started about a million years earlier. It lasted for 4 million years, extending into the beginning of the Silurian Period. Initially, the changes likely appeared as minor perturbations in the environmental status quo, but by the end, 85 percent of Ordovician species had vanished into the void of extinction. Perhaps there is a lesson for Anthropocene climate change deniers and agnostics in this ancient event.

Four million years is a long time, and the progression of extinction and subsequent recovery occurred in pulses. The earliest casualties appeared as many of the graptolites, brachiopods, and trilobites faded away. Dropping seal levels signaled the onset of phase two as water drained from the shallow seas, destroying the cradle of Ordovician marine life. The final phase commenced during the early Silurian, when global cooling transitioned to a warming trend, and sea levels began to rise once again.

This long, drawn-out extinction differs from mass extinctions caused by discrete events such as flood basalts or asteroid strikes. It's worth remembering that mass extinctions come in a variety of flavors. Several billion years before the Ordovician-Silurian mass extinction, the introduction of oxygen into Earth's biosphere wiped out large portions of chemosynthetic life. Then, during the Cryogenian crisis, it was just 'too damn cold' to live. The absence of oxygen proved deadly to life at the end of the Cambrian. We will see later in this story that some extinctions have been triggered by massive volcanic eruptions, which brought on rapid environmental change and disrupted life. Lastly, we are all familiar with the fate of the dinosaurs and should not forget the possibility of extinction by an asteroid strike.

Cooling Off

Rising glaciers at the South Pole on the Gondwana supercontinent locked away needed water as ice, and at the same time, Ordovician plants facilitated the depletion of carbon dioxide in Earth's atmosphere. Steady cooling as CO_2 levels dropped encouraged even more glaciation. Earth was locked into a feedback loop, the perfect storm to plunge the planet into a cold spell.

We don't know exactly how cold the planet was. Paleoclimate investigations suggest that late Ordovician atmospheric and oceanic temperatures ranged from slightly warmer to slightly cooler than today. However, it is important to remember the geographic context of this period. Most landmasses were in the southern hemisphere, leaving the northern hemisphere as an open ocean.

The massive Gondwana supercontinent provided ample room for the largest ice sheets of the Paleozoic. Some researchers estimate that glaciation at the end of the Ordovician covered an area twice the size of the most recent Pleistocene ice age glaciers, and the Ordovician ice sheets were twice as thick as those of the Pleistocene.

A simple geometric model of these numbers says the Ordovician glaciation may have locked up four times as much water as our last ice age. Global sea levels were 120 meters lower during maximum Pleistocene glaciation, so they may have dropped by 480 meters at the end of the Ordovician, even without accounting for thermal contraction.

Ocean temperatures were dropping at the end of the Ordovician. The same physics of thermal expansion, which added to sea-level rise as the mid-Ordovician world warmed and ocean volume expanded, worked in reverse. Cooling oceans caused water volume shrinkage, further exacerbating sea-level drop.

Death by Air

Debates about the exact causes of the Ordovician-Silurian mass extinction are varied and speculative. As always, science is largely determined by the weight of evidence, and science is often a world of competing theories, not absolute facts. Individual scientists are like cautious gamblers weighing all the evidence and placing their bet for a winner on the weight of supporting evidence.

I will roll the dice and mark up this particular mass extinction as a loss of habitat. Toxicity doesn't appear to be an issue, although some researchers have proposed that an increase in toxic metals may have killed off the base of the Ordovician food chain. Likewise, the cold alone doesn't appear severe enough to be the major triggering factor. It did get colder than the mid-Ordovician, but with estimated temperatures close to today and a gradually cooling environment, it seems that evolutionary change would have accommodated the global temperature drop.

We lack any evidence of massive volcanic and flood basalt activity at the end of the Ordovician Period. An asteroid strike can't be ruled out, and there is some evidence of a massive impact crater in Australia that may have originated in the late Ordovician, but the jury is out on the exact timing of this event. Loss of habitat remains a prime candidate as the cause of the Ordovician-Silurian mass extinction. Mid-Ordovician life thrived in vast shallow seas, and during this period, Earth's oceans rose to the highest levels in geological history. Large portions of Earth's continents were submerged beneath shallow seas, and Ordovician life depended on this unique environmental combination of shallow water, ample sunlight, and effective oxygen exchange.

Gondwana glaciation and cooling seas resulted in a devastating fall in ocean levels. Water drained from the continents. Shallow seas became dry land, and all of life there experienced death by air because most life outside the wet, briny womb of the oceans was not yet possible. There were no more vast shallow seas, only waves lapping at the edges of steep continental slopes, and so began the Silurian Period.

A Reversal of Fortune

The Ordovician-Silurian mass extinction was one of the biosphere's historic low points, so the Silurian Period came in on a sour note. When the Silurian recovery started 443.8 million years ago, only 15 percent of the species present during the Ordovician remained. Fortunately, the tide turned for evolution, and life was on the upswing. The ecological void left by a catastrophic mass extinction provided evolution with plenty of opportunities to innovate.

At the start of the Silurian, the Gondwana supercontinent still covered much of the southern hemisphere in and around the South Pole. Still, warmth slowly returned to the planet, causing the magnificent southern ice sheet to shrink. Rising temperatures and melting ice led to a significant sea-level rise.

The causes behind this reversal of fortune are uncertain. Changes in solar radiation levels are possible but undocumented. One suggestion points towards changes in the influx of galactic cosmic rays from nearby supernovae, but this also seems to be on the fringe of possible explanations.

In many warm periods, a historical culprit for rising global temperatures is the degassing from massive flood basalt eruptions. Large quantities of carbon dioxide enter the atmosphere during these events, creating a greenhouse effect. But a search of recognized flood basalts places the earliest of these events at about 377 million years ago, long after the end of the Silurian at 419.2 million years.

Perhaps the massive chemical weathering from the initial invasion of continental landmasses by plants had run its course, reducing carbon dioxide uptake from weathering. Such a change may have led to a net accumulation of CO_2 in the atmosphere. Silurian carbon dioxide levels were higher than our modern levels, with estimates ranging from three to ten times higher.

Despite the causes of warming, life benefited from Silurian environmental conditions and thrived again in the warm shallow seas created by rising oceans. Evolutionary progress was not limited to the oceans, and significant changes also occurred on land.

Life at Sea

Mother nature was busy during the Silurian, building the foundation of ichthyology as fish rapidly evolved. Jawless fish were present at the beginning of the Silurian, and they continued to thrive, expanding their range and rapidly spreading around the planet. But fish evolution took two big steps forward during this period. The first freshwater fish appeared, and evolution equipped some newly evolved species with jaws. What better way to secure your dinner than clamping down on it so it can't escape? This evolutionary adaptation was clearly an important step towards the production of the famous sea-horror movie 'Jaws.'

But fish were not at the top of the food chain. Eurypterids, the apex predator of the Silurian shallow seas, held this hallowed position. These arthropods were like sea scorpions on steroids with jointed legs for walking, pinchers for grabbing, paddle-shaped appendages for swimming, and a tail with a venomous spike at the end. These sea monsters grew up to eight feet long, making them the largest arthropods to ever live on Earth.

Despite clear evolutionary progress during the Silurian, animal life remained largely confined to the oceans. A stunning array of marine species lived in the Silurian shallow seas. Just as in the Ordovician, higher sea levels allowed the oceans to spread over vast areas of low-lying continental topography, creating the perfect habitat for ancient marine life.

Atmospheric oxygen levels during the Silurian were lower than today, probably hovering between 35% and 65% of modern levels. Also, the climate was warmer by about five degrees Celsius. Warm water holds less oxygen. So, between lower atmospheric oxygen and a reduced capacity for absorbed oxygen in the seawater, life was generally limited to a depth of about 100 meters below the ocean surface. Waters deeper than 100 meters were often anoxic.

Life on Land

While life at sea was rapidly evolving, changes were also occurring along the shorelines. Life continued to creep out of the ocean as vascular plants and the first macroscopic animal life appeared. Land plants provided food, and the precursors of spiders and millipedes evolved to take advantage of the new food chain, scavenging these first shoreline ecosystems, eating plants and decaying detritus, and preying on each other.

The mosses and hornworts of the Ordovician were joined by a more robust form of plant life that would eventually come to dominate our terrestrial ecosystems. Vascular plants gained a foothold and utilized their morphological resources to grow upward and outward, enabling them to more effectively harness the free energy from the sun.

Earth's restless tectonic plates were in motion, and a mountain-building episode that began in the Ordovician intensified during the Silurian, resulting in the formation of the Scottish Highlands and associated mountain belts in Scandinavia and Greenland. The landmass of Laurentia formed when North America, Baltica, and Avalonia (Western Europe) collided in a massive orogenic event that sutured together the three plates. The Iapetus Ocean, which originally separated these three continents, disappeared as the plates merged together into a new continent.

This episode was part of the Caledonian Orogeny, an event describing a series of continental plate collisions occurring over a 100 to 200-million-year period, depending on which researchers you speak with. So, the Silurian collision was only part of a much larger and longer process.

Overall, the Silurian Period was a time of rapid evolutionary change.

Earth's ecosystems were maturing and actively shifting onto dry land. The food webs and energy chains of our modern biosphere gained a significant foothold during the Silurian. The evolutionary progress on land was particularly significant, as it laid the foundations for a terrestrial biosphere that would lead to widespread forestation across the globe.

Devonian Trees Change the World

The tree of life towers above with roots deep in the ground, connecting heaven, earth and a mysterious underworld.

An axis mundi in the mind of the shaman.

Passage into the Devonian

The Devonian Period spans 60 million years, from 419.2 to 358.9 million years ago. Earth was hot, and sea levels were high. Today, oceans cover 71 percent of Earth's surface, but during the Devonian, the planet was a blue jewel circling the sun with 85 percent of its surface underwater. Shallow seas abounded, creating vast marine ecosystems for life to exploit. The period is sometimes called the Age of Fishes. Our water-bound friends were the apex of animal life as evolution created a host of new marine species. But while fish were ruling the planet, some four-legged vertebrates finally snuck from the ocean, virtually unnoticed, and crawled out onto dry land.

The end of the Silurian and the beginning of the Devonian Period was mostly a footnote in the annals of geological history. Extinctions in the fossil record separate all major periods of the Paleozoic, but these extinction events come in all shapes and sizes. Species loss at the end of the Silurian was minor compared with many other extinctions and probably caused by changes in the climate and the environment. There was death but not outright devastation.

Mother nature sometimes takes on the specter of a mad, passionate artist, wiping the canvas clean out of frustration and painting a new creation. At other times, she's gentler, smudging over and repainting portions of the canvas but leaving the gross composition intact. The change from Silurian to Devonian was the latter of the two. But during the Devonian, she painted a dizzying array of new fish onto the canvas.

Beneath the Sea

Most animal life during the Devonian remained in the shallow seas and coastal waters, so evolution worked with what was available. Armored fish, along with jawed fish, flourished. The two seem to go hand in hand. Once your neighbors develop jaws, you surely want some armor to avoid being clamped down on for dinner. Scaly armor on the outside keeps the insides where they belong.

Sometime around the middle of the Devonian, a group of fish arose called Chondrichthyes. These were precursors to sharks with a cartilaginous skeletal framework. Today, this class of fish also includes skates, rays, and chimeras.

Nature was also busy perfecting the use of coiled shells in the Early Devonian, and a variety of ammonites appeared. Similar to today's nautilus, these ancient predators used their coiled shells for protection and mobility. A tube called the siphuncle runs through the chambers of the nautilus' coiled shell, allowing the release of fluids and gases to maintain neutral buoyancy and float in the water column. The siphuncle in a nautilus runs through the center of its shell chambers. But the ammonites' tube ran along the outer edge of its shell. I speak of them in the past tense because ammonites are no longer with us.

But perhaps the most interesting development in the Devonian fish world occurred within a now-extinct group of fish called Rhipidistia. Fossil remains indicate they were bony fish with a lobe-finned anatomy. Their fins extended out from two distinct pairs of ventral and dorsal lobes. Not too much imagination is needed to extrapolate these lobes into budding legs. Detailed analysis of their fossil remains supports the theory that these fish evolved into the first four-legged tetrapods to venture onto land.

Life on the Move

In parallel to the Rhipidistia, lungfish also appeared in the Devonian. Their evolutionary trick of living for extended periods out of the water and surviving on atmospheric oxygen offered another part of the solution for land-based animal life. Insects may have been the first animals on land during the Silurian, but four-legged life was not far behind. By the end of the Devonian, animals gained a strong foothold on Earth's continents. The earliest of these tetrapod invaders were probably salamander-like amphibians.

Earth's continents were on the move also. The mighty supercontinent of Pangea was consolidating as the two main landmasses, Laurasia and Gondwanaland, moved closer together, and the last remnants of the Iapetus Ocean disappeared. The collision of these two massive continents resulted in large-scale mountain building, producing the early Appalachian Mountains and continuing the rise of the Caledonides in Britain and Scandinavia.

The success of life in the Devonian resulted from the interplay of climate, geology, and biology. A warm climate and associated high sea levels provided long-lasting shallow seas. Over 6,000 feet of limestones and shales were deposited in what is now Utah, and in Europe extensive marine deposits covered areas from England to Russia. The period also saw some of the most prolific coral reef growth in geological history.

Animal life on the planet surged ahead, but plants also made great strides. Vascular plants and the first true trees appeared on the evolutionary stage. The ramifications of this explosion in plant life may have been more significant for Earth's history than the changes wrought in the animal kingdom.

A Revolution of Vascular Plants

Trees, structured like the ones we know today, didn't appear on the scene until about 385 million years ago during the middle Devonian—deep roots, a stout bark-covered trunk, and branches supporting a thick cover of green leaves. They probably didn't look much different from the trees in your backyard or a city park.

The rise of vascular plants in the preceding Silurian Period was an evolutionary breakthrough, and it formed the beginning of a revolution on land. The ability to effectively transport fluids and nutrients from the ground to a plant's distant extremities created a rich fabric of opportunity, allowing these vascular plants to exploit new niches in Earth's emerging land-based ecosystems.

Trees evolved during a time when the climate was warm and relatively stable. Plant evolution experimented with various strategies for surviving away from bogs and open water. Competition for resources sparked new approaches, simultaneously sending plants higher above ground and deeper below the surface.

Competition to grow tall and wide was fierce because access to direct sunlight determined who captured the most energy. However, extra height induced structural instability, making it difficult for the plants to stay upright. This problem with the physics of gravity facilitated growing both upward and downward, developing deep root systems to keep the growing mass above the ground from toppling over. These deep roots allowed trees to excel at capturing water and nutrients deep below the surface.

Trees

The earliest trees developed survival strategies that were so successful they are still in use 385 million years later. The rise of trees in the middle Devonian was a turning point in Earth's history. Terrestrial ecology was forever changed, along with climate and geochemical cycles—trees became a major force shaping the surface of our planet.

Investigations into paleosol (fossil soil) in the New York Catskill mountains recently uncovered definitive evidence of fossilized root systems from the genus Archaeopteris, Earth's earliest-known true tree. The preserved evidence shows a highly evolved root system similar to modern-day trees.

Anyone who has flown over the vast boreal forests of the Northern Hemisphere, hiked the Appalachian Trail, visited the Brazilian Rainforest, or driven through the coastal range of Oregon knows trees are the backbone of woodland ecosystems. These types of forests dominate large areas of our planet.

Trees are the anchor species around which entire ecosystems survive. We have learned the hard way: destroy the trees, and you destroy the ecosystem. The trees of ancient Devonian forests paved the way for rich and diverse terrestrial ecosystems.

Trees alter the environment by digging into the ground and breaking down rock into soil. During the process, nutrients are released for a variety of living organisms to utilize. Water and erosion carry these nutrients to the oceans, where they boost marine productivity. Trees remove organic carbon from the air, storing it in their biomass and in the soil below. This process allows the development of humus-rich soils—a perfect setting for plants, bacteria, and animals to flourish.

Earth's new Devonian forests altered the climate and landscape, creating new dynamics for the world's food webs and providing vast new ecological niches for evolution to fill.

The Tree Effect

The impact of trees on Earth's history speaks to how they interact with rock, soil, and air. Deep roots allow trees to physically and chemically alter Earth's surface. As the roots break down solid rock, they expose more surface area to weathering. Weathering involves both physical and chemical processes. The physical process produces smaller particles and allows for increased erosion as rainwater carries these particles away into rivers and streams. Chemical weathering breaks down the mineral components of the rock, releasing nutrients. Importantly, chemical weathering also absorbs carbon dioxide from the atmosphere, altering the balance of greenhouse gases. Less carbon dioxide in the atmosphere leads to a cooler climate.

Another aspect of a more forested Earth is the effect of trees on marine food webs. A slow, steady influx of nutrients into coastal oceans enables higher marine productivity and more robust ocean ecosystems. However, if forestation results in excessive nutrients entering the waterways, then algae blooms form in the coastal oceans. Massive algae blooms release toxic compounds and cause anoxia by using all available oxygen in the water during periods of hyper-productivity.

Rapid forestation of Earth's continents during the Devonian came at a price. It disturbed the existing balance between the biosphere, atmosphere, geosphere, and hydrosphere. The final effects of these changes heralded the end of the Devonian.

A Devonian Disaster

All actions influence the future and what is done cannot be undone. Some actions unleash a chain of events we can not escape from. A small disaster today may open Pandora's box, leading to catastrophe tomorrow.

Despite life's advances on land and at sea during the Devonian, the circle of time completed a loop. Between about 383 and 359 million years ago, 70 to 80 percent of the species on our planet disappeared. However, evidence defining the cause of this catastrophe is sparse. Perhaps life was too successful for its own good. After all, the ongoing Anthropocene extinction provides strong evidence that human beings may be too smart for their own good.

It's important to place this extinction event into the proper time frame. The Late Devonian is divided into two stages: the Frasnian followed by the Famennian. These two stages span 24 million years. Paleontologists identify three periods in the Late Devonian during which major changes occurred in the fossil record: the beginning and end of the Frasnian and the end of the Famennian. Large numbers of species disappeared from the geological records at these points in time.

The Late Devonian extinction didn't occur as a single catastrophic blow to life. It was more like a slow-moving train wreck or death by a thousand cuts. So, it seems doubtful an asteroid strike was the primary culprit. Impact craters like the Siljan Ring (Sweden), Alamo (USA, Nevada), and Woodleigh (Western Australia) all are roughly dated to the Late Devonian, but none of these craters are the size of the Chicxulub impact that caused the Late Cretaceous extinction. Also, the dating of these events is not precise enough to definitively tie them to any of the three extinction peaks.

Perhaps the 24-million-year interval encompassing the Late Devonian extinction represents a gradual development of severe chemical and environmental imbalances. Simply put, Devonian habitats became inhospitable to life.

Organic Shales

The term '*shale*' refers to fine-grained sedimentary rocks composed mainly of silt- and clay-sized particles. Usually, these rocks originate from sediments deposited in waters away from currents and surface disturbances. The deposits then build layer upon layer over thousands to millions of years.

However, there is a particular type of shale highly sought after by the oil exploration industry, known as black shale. Its dark color comes from the high levels of organic matter trapped in the rock matrix. When black shales are buried deep in the ground, and temperatures rise to between 110 and 120 degrees Celsius, the organic matter in black shales begins converting to oil. Once it starts generating oil, it becomes a '*source rock.*' Follow the source rock, and you find the oil.

The organic material in black shales is an indicator of ancient environmental conditions. Normally, organic materials like the remains of dead plants and animals fall to the ocean floor, where bacteria consume them. However, when the deep ocean waters are anoxic and devoid of oxygen, there are no ocean-bottom organisms to consume the organic matter raining down from above, and it becomes incorporated into the sediments. The presence of black shales indicates persistent anoxia, and we know that the absence of free oxygen means the absence of most life.

Vast deposits of Devonian black shales cover much of the eastern United States, indicating widespread anoxia in the Late Devonian oceans. These shales point to severe ecological stress since most animal life still lived in shallow marine ecosystems and depended on oxygen for survival.

Late Devonian Anoxia

A variety of physical and environmental conditions cause ocean anoxia, and it is important to understand the potential sources of oxygen depletion. Today, we have widespread anoxia in the Black Sea below about 100 meters. This anoxia results from density stratification between deep, cold, saline water and a thin, shallow surface zone with fresher water. The density difference is high enough that the two don't mix. Because ocean water requires contact with the atmosphere to oxygenate, no oxygen reaches the denser layer below the surface water. In this case, the physiographic conditions of the Black Sea basin led to its permanent anoxia.

However, anoxia can also occur in bodies of water without stratification. The process is biological and starts at the bottom of the food chain. Organic, nutrient-rich waters provide a paradise for algae, so algae populations explode when excess nutrients enter their ecosystem. As the algae die, bacteria consume the organic remains. But the bacteria have metabolic needs and require oxygen. So, when excessive dead organic matter appears, bacterial activity increases and slowly depletes the free oxygen.

The Devonian Period witnessed the rapid evolution of plants as they grew larger and developed increasingly complex root systems. During the Devonian period, the plants and trees that dominated dry land played a crucial role in breaking down rock into soil. The geological record indicates widespread forests developed during the Late Devonian. Under these conditions, the amount of organic material transported by streams and rivers to the oceans would have rapidly expanded. Also, increased chemical weathering due to forestation provided more nutrients to marine habitats.

Conditions were ripe in the Late Devonian for persistent anoxia in the extensive shallow oceans that were still the cradle of the animal kingdom. A perfect recipe for mass extinction.

Ozone

Life is finicky. A single oxygen molecule doesn't do the trick for efficient cellular respiration. But put two of these molecules together, and you have free oxygen, an essential ingredient for life in Earth's biosphere. Two is the magic number. Adding a third oxygen molecule creates a gas called ozone, which accumulates in the stratosphere between 7 to 30 miles above Earth's surface.

The positioning of the ozone layer is fortuitous for humans since high concentrations of the triple oxygen molecule are harmful to our lungs. But the ozone layer serves another life-preserving purpose. Stratospheric ozone forms from the interaction between free oxygen (O2) and solar ultraviolet (UV) radiation, and ultimately, the ozone layer protects the biosphere from the Sun's harmful UV radiation.

Without the ozone layer, both plant and animal life would suffer from the effects of excessive UV radiation. One of the nastier consequences of increased UV exposure is higher rates of genetic damage.

Researchers at the University of Southampton discovered something interesting while studying pollen cell walls from the Late Devonian period. The samples used in their study came from rocks in East Greenland that span the boundary between the Devonian and the subsequent Carboniferous Period. Their work revealed a high incidence of genetic damage in plant spores at the end of the Devonian period, approximately 359 million years ago. The damage they observed was consistent with deformities caused by excessive exposure to UV radiation.

Radiation Poisoning

Plant spore DNA is particularly sensitive to UV radiation during the plant's developmental period before its protective sporopollenin walls form. Excessive radiation at this phase of their life cycle causes genetic damage.

The East Greenland samples provided interesting insights into life at the end of the Devonian. Increasingly damaged spores preceded a notable decline in spore diversity, indicating the extinction of many plant species. These events were tied also directly to the disappearance of large, stemmed plants in the overlying rock strata.

The researchers theorized that the end of the Devonian coincided with a loss of Earth's ozone layer, and the resulting genetic damage caused a collapse of the Late Devonian forest ecosystems.

The cause of the Late Devonian ozone problem is postulated to be rising continental temperatures, which allowed more convective water transport into the upper atmosphere. The water vapor carried chemical compounds, including chlorine, which is produced naturally by some plants and algae. When the chlorine reached the stratosphere, it broke down the ozone, thus allowing damaging ultraviolet radiation to reach the Earth's surface. Essentially, the problem was a weakened ozone layer caused by climate warming.

The last stages of the Devonian coincided with the onset of glaciation. However, the glacial episode was short-lived, followed by warming that extended into the early Carboniferous, where estimated global temperatures were 20 degrees Celsius, significantly warmer than today. This scenario is consistent with the theory that rising continental temperatures may have affected Late Devonian ozone levels.

All of the Above—Cascading Events

We know mass extinctions are caused by rapidly changing environmental conditions. Still, because of the lengthy 24-million-year decline encompassed by the Late Devonian extinction, its specific causes are more speculation than fact. But there are various bits of evidence hinting at possible causes, and a possible answer is all of the above—the perfect storm.

A classic perfect storm happened in 1991, and the movie version was produced in 2000. The meteorological backstory details a set of unusual circumstances that allowed a nor'easter to absorb a hurricane and evolve into a superstorm. This storm resulted in thirteen deaths and thirty-foot waves pounding the east coast of the USA, causing approximately half a billion dollars of damage in today's money. Cascading natural events led to the perfect storm.

Perhaps the end of the Devonian was a combination of cascading global events. The expansive growth of continental forests caused frequent over-fertilization in the oceans and subsequent widespread marine anoxia. The combination of plant and algae growth also pumped more natural chlorine into the atmosphere, weakening the ozone layer. The global collapse of the Devonian forest ecosystems due to ozone depletion then led to a massive influx of organic material into the oceans, further prolonging the anoxia.

We don't know the answer, but 70–80 percent of Earth's species disappeared. Extinction events are often associated with dramatic climate change, and the Anthropocene extinction is no exception. We cannot see all of the environmental connections, and a series of cascading events can lead to unexpected disasters. Rising CO2 levels cause higher temperatures, leading to Arctic thawing, Antarctic melting, rising sea levels, and ocean acidification that kills off the base of the marine food web. But we can already see this chain of events that is happening today. What is out there that we can't yet see?

Carboniferous Period: The Age of Coal

Within the foundations of our success always lie the seeds of our destruction.

It is inescapable that the scales of evolution depend on the mutual interaction of success and failure to stay in balance.

Setting the Stage

The Carboniferous Period derives its name from our love of fossil fuels. Between 359 and 299 million years ago, life laid the foundations for Anthropocene climate change as massive coal deposits formed. Ultimately, these ancient coal deposits provided the energy that fueled the Industrial Revolution, which began in the mid-1800s. The world learned to extract cheap energy from the Earth's bowels and convert it into work, driving the development of industry, manufacturing, and transportation. It would be another century before humans understood that this '*cheap energy*' came at a high price, and even longer before we grasped the long-term environmental and climate implications. Oddly enough, the creation of the fossil fuels that now warm our planet resulted in catastrophic cooling at the end of the Carboniferous Period.

European geologists recognize the Carboniferous Period as a single geological interval. However, North American geologists treat the Carboniferous as two distinct periods, with the Early Carboniferous designated as the Mississippian Period and the Late Carboniferous referred to as the Pennsylvanian Period. These divisions relate to the prevailing environmental conditions during each period. Large, shallow continental seas dominated the Early Carboniferous, whereas the Late Carboniferous was marked by more dry land.

During the Carboniferous, Gondwana finished migrating northward toward the equator and fully fused with Laurasia completing the supercontinent of Pangea. The tectonic collisions associated with the formation of Pangea resulted in extensive uplift and mountain building. This orogenic activity created a supercontinent with glaciated highlands and interior lowlands where the surrounding mountains trapped plenty of warm, moist air. Weathering and erosion of the highlands carried sediments, minerals, and nutrients to the lowlands, making them rich and fertile. In these swampy lowlands, fertile soil and a warm, moist climate sparked the growth of vast forests.

As plants and trees died and sank into the swampy muck of these forest floors, they were quickly buried in anoxic bogs before bacteria could devoir their remains. Over millions of years, these deposits evolved into thick layers of carbon-rich sediments, which were ultimately pushed deep underground, where heat and pressure transformed them into coal deposits. The vast majority of Earth's coal deposits (up to 90%) were formed during the Carboniferous period.

Like a double-edged sword, the availability of this late Carboniferous coal cuts two ways. It simultaneously set the stage for technological innovation and the success of the Industrial Revolution, as well as the subsequent rapid environmental changes over the next 150 years. The coal provided cheap energy, driving industrial expansion and the development of modern technology. It also initiated climate warming at a rate not seen in our geological history.

Carbon and Oxygen

During the Carboniferous Period, the atmosphere was warm. As plant life and forests recovered from the Devonian extinction disaster, the continents again became lush and green. Vascular plants dominated Earth's terrestrial ecosystems, allowing plants to grow taller and broader, thus increasing their impact on the environment.

Giant horsetails, ferns and fern-like trees, primitive conifers called cordaites, and a host of other plants carpeted the landscape doing what they do best, sucking carbon dioxide from the atmosphere and producing oxygen.

The success of terrestrial life during the Carboniferous Period was stunning when measured by the amount of organic carbon it added to the geological record. Trees and other plants perfected the art of capturing sunlight and turning it into food for growth. As photosynthesis went into overdrive, Earth's forests poured newly created oxygen into the atmosphere. Oxygen was being produced faster than it was consumed, heralding an unprecedented rise in atmospheric oxygen levels. Near the end of this period, oxygen levels rose to 35 percent, a 75 percent increase over modern-day levels.

Higher oxygen levels created a denser atmosphere and intensely blue skies. But this oxygen also affected the plants and animals in Earth's terrestrial ecosystems.

Insects and Fire

Anthropocene climate warming is increasing the frequency and intensity of wildfires. Warmer and dryer conditions around the globe have created massive forest fires in the boreal forest of Siberia and propagated devastating fires in Australia, ravaging large areas of the country. Each year fires in California and the Pacific Northwest become more intense, sparking fierce firenadoes and burning whole communities to the ground.

It was not always this way on Earth, and for most of the planet's history, wildfires were unknown because atmospheric oxygen levels were too low. At oxygen levels of 16 percent or lower, wildfires won't propagate, and at levels of 23 percent or higher, destructively intense fires spread uncontrollably. So, during the Carboniferous period, with atmospheric oxygen levels of 35 percent, Earth often degenerated into a burning hell.

Increasing fire frequency is correlated to periods during Earth's history when high oxygen levels prevailed. Evidence of frequent fires in the Carboniferous is found in fossil records, which track the occurrence of charcoal from wildfires.

Animal life in the Carboniferous Period was also affected by high oxygen levels. Some forms of life benefit greatly from extra oxygen, and the Late Carboniferous marked the reign of giant insects. Like all animals, insects need oxygen for cellular respiration. But insects don't use lungs and a circulatory system to distribute needed oxygen. Instead, they have a network of tiny tubes distributed throughout their bodies, allowing oxygen to reach all necessary tissues. It is a relatively inefficient way to distribute oxygen, and it limits their ultimate size since tubes that are too long run out of oxygen before reaching their final destination.

However, air rich in oxygen allows the insects to grow larger. The predatory Carboniferous griffin flies, dragonfly-like creatures, had wingspans of over two feet. Giant carnivorous dragonflies, I don't like the sound of that.

The Amniote Egg

Once life crawled out of the ocean, it sought to explore a strange new world and go where no life had gone before. However, before such a journey was possible, life had to solve some basic problems associated with living on dry land. One such pressing matter was how to reproduce. Sexual reproduction was nothing new, but that left the problem of the egg.

The Carboniferous Period produced one of life's great evolutionary innovations, the amniote egg. The movement of animal life onto land created a reproductive dilemma. Prior to the first land settlers, animals lived beneath the waves of ancient oceans. There, they reproduced in a wet, aqueous environment where fertilized eggs developed into new life. But this strategy had its limitations on land. Perhaps these water-bound eggs might survive in damp, low-lying terrains, but they were susceptible to desiccation as climate conditions changed and periods of drought arrived. These eggs stood little chance of survival away from the water's edge.

Amniote eggs changed this dissatisfactory state of affairs by providing a tough membrane seal around the egg, preventing it from drying out. These were probably rubbery textured membranes at first, but eventually, they led to hard-shell eggs. You will notice there were eggs but no chickens in the Carboniferous, so here we find the answer to the age-old riddle of which came first: the chicken or the egg? This innovation allowed the ancestors of reptiles, birds, and mammals to exploit ecological niches across the continents and far from the oceans.

Amniote eggs encase their embryonic life in a sealed capsule with all the fluids and food needed for a new life to develop. The liquid ingredients for life stay safely inside, yet air, hence oxygen necessary for life, can still permeate the shell. All in all, it was a very clever evolutionary development.

Self-Regulating Cycles

The biosphere and our environment are in a constant dance of self-regulation. Feedback loops of all types adjust and modulate environmental conditions. The end of the Carboniferous Period is a classic example of a self-regulating cycle.

The period was a remarkable time for both animal and plant evolution. Conditions were ripe for the growth of thick forests across extensive areas of low, swampy terrain. Ample sunlight, water, and nutrients allowed for explosive growth. The most productive areas were in the equatorial Laurasia landmass, where grand Carboniferous rainforests developed. Geological conditions conspired to create environments where the rapid burial of dead trees and plants allowed for thick, rich coal deposits to develop. Over 300 million years later, these same coal deposits drove our planet's Industrial Revolution and kicked off Anthropocene climate warming.

Coal is just condensed carbon. Trees and other plants extract carbon dioxide from the atmosphere and transform it into organic carbon. During the Carboniferous, this organic carbon was sequestered in biomass and in the forest soils. When plants died, carbon locked in nascent coal deposits was permanently removed from the atmosphere.

The late Carboniferous, when many of the coal beds were deposited, spanned a period of 24 million years. Slowly the amount of carbon dioxide in the atmosphere decreased as the forests sequestered it, leading to global cooling. Cooler and dryer conditions eventually caused the collapse of the Carboniferous rainforests, followed by increased high-latitude glaciation.

Success always comes at a price because it changes the environmental balance. The tremendous success of the Carboniferous rainforests depended on a warm climate and lots of carbon dioxide. Eventually, the rainforests significantly depleted the atmospheric carbon dioxide. They became the unwitting designers of their own demise.

A Long Time Coming

The environmental saga of the Carboniferous stretched over tens of millions of years. Despite the severe environmental changes, the period ended with only a minor extinction event, centered mainly around rainforest collapse in the tropical latitudes. Environmental shifts over long periods of time provide a buffer for the biosphere, giving evolution a chance to work out solutions as ecosystems slowly change. Average global temperatures shifted from about 20 degrees Celsius in the Early Carboniferous to about 10 degrees Celsius by the end of the period.

The average rate of temperature change during the Carboniferous was about 0.167 degrees Celsius per million years or six million years for a temperature drop of one degree. We have seen an Anthropocene temperature rise of one degree in 100 years by comparison. The comparative rate of change between now and then is staggering, with Anthropocene change occurring thousands of times faster than during the Carboniferous period.

The rate of change matters; the faster the rate of environmental change, the less likely it is that evolution can keep up. The end result of this imbalance is species extinction, with increasing species loss as the rate of change accelerates. Homo sapiens are clever, but they cannot escape the fallout of rapid climate change. This fallout does not guarantee extinction for a species as adaptable as humans, but life would look very different if our critical marine species suffered a massive population decline. Over three billion people currently depend on seafood as their primary source of protein.

The industrial and technological revolutions of the Anthropocene mean that human activities have profound effects on the ecosystems that sustain us. It is entirely possible for us to destroy the ecosystems we ultimately depend on for survival and become the masters of our own demise.

The Great Dying

The ebb and flow of life on Earth highlights the biosphere's precarious balance between life and death.

The biosphere exists at the triple junction between the geosphere, atmosphere, and hydrosphere.

The Heyday of Pangea

Circumnavigating the globe by ship is an arduous affair. Like the legendary Ferdinand Magellan, if you start in Europe and head west, you eventually bump into the Americas. A turn to the north will be futile due to Arctic ice, so you head south. Passing by the Caribbean and Central America, you wind your way southward along the eastern coastlines of Brazil and Argentina. Finally, you sneak around Cape Horn, braving the rough and windy seas of the Southern Ocean, enter into the Pacific Ocean, and head northwest.

Life moves along leisurely until you brush against Australia and Indonesia, maybe even detouring north to China. As you seek your return to Europe, you probably visit India and the Middle East, and eventually, you traverse the length of Africa, zip around the Cape of Good Hope, and make the long journey north to your home port. It's quite the journey to sail around the globe, with your trip constantly interrupted by large, elongated landmasses.

If you attempted the same journey 270 million years ago during the Permian, you would have taken a simpler path. You would have sailed westward from an equatorial port on the west coast of Pangea and traveled in a straight line until you arrived back on the continent's east coast. During the Permian, 70 percent of Earth's surface was covered by water, forming a single continuous ocean. The remaining 30 percent of the surface, covered by dry land, was glued together as our favorite supercontinent, Pangea.

North and South America occupied the equatorial and mid-latitude areas of Pangea's west coast, and Africa and the Middle East were positioned on the equatorial east coast of the continent.

The Permian Period marked the end of the Paleozoic. It occupied a 47 million-year stretch from 299 to 252 million years ago—just a slightly shorter interval than the transition from dinosaurs to smart ape-like creatures called Homo sapiens.

The superfamily Hominoidea popped onto the evolutionary scene about 20 million years ago, and a little over 19 million years later, Homo sapiens took center stage. So, the Permian Period was twice as long as our entire evolutionary history. No species as intelligent and deliberately negligent as Homo sapiens developed during the Permian, but still, it was quite an important time for animal life in general.

Reptiles came into their own during the Permian, and important lineages developed. Permian protorosaurs were the ancestors of archosaurs, a group that eventually developed into dinosaurs and birds. Closer to home for you and me, the synapsids, reptiles with some mammalian traits, evolved during the Permian. These Permian creatures set the stage for true mammals to develop in the Mesozoic.

The Permian world was ripe for Mother Nature to experiment. When she discovered ingenious evolutionary innovations, there was plenty of connected land, free from deep ocean barriers, for the biosphere to spread its success across vast and diverse ecosystems.

The Permian plate tectonic configuration was not quite as simple as initially described above, but there was a considerable amount of mid-continent territory to contend with. These mid-continent areas were home to glaciers at higher altitudes and desert drylands when the mountains blocked moist air from the oceans.

Climate Helps

Global cooling at the end of the preceding Carboniferous Period meant the Permian started on a cold note. Pangea was truly a supercontinent, and its lands stretched from the South Pole almost to the North Pole. The ancient Gondwana landmass occupied the southern part of Pangea, providing a vast continental interior from the South Pole to the equator. This configuration promoted the formation of large glacial ice sheets, which persisted from the Carboniferous into the Early Permian.

But the planet eventually warmed, and warm moist climates dominated many areas by the mid-Permian. The climatic conditions became ideal for life to blossom and thrive in both temperate and higher-latitude ecosystems. Because Pangea created a formidable north-south land barrier, it affected the flow of ocean currents, deflecting warm waters from the tropics to higher latitudes. This led to milder climates near the north and south polar regions. Perhaps having the same effect as the modern Gulf Stream, which warms the North Atlantic, creating milder climates in Europe than would normally be expected.

But change is the only constant, and by the Late Permian, warm had turned into hot, and desert conditions became widespread. Multiple factors were at play, but geography was a major influencer of local climatic conditions. Pangea was the child of multiple orogenic episodes as tectonic plates collided and extensive mountain belts rose skyward. These mountains affected air currents and weather systems, sometimes cutting off moisture to large inland basins.

Today, ancient Permian rock strata are a major source of evaporite minerals like potash salts. These deposits formed under extremely hot, dry conditions, and they were particularly prevalent in the tropical and subtropical regions of Pangea. The Permian also gifted us lots of oil—a blessing and a curse as we simultaneously satisfy our need for cheap power and radically alter our climate and biosphere.

The End of Early Life ... Almost

The Permian Period came to a catastrophic end 252 million years ago with the extinction of 95 percent of all marine species and 70 percent of terrestrial species. Most of life checked out at the backdoor and disappeared into the ether in an event known as the Great Dying.

Humans are not great planners. Individually we do reasonably well, but as a group, we are prone to procrastination. This flaw is why politics is mostly about reactive policy rather than proactive planning. Why do today what we can put off until it becomes a full-blown emergency?

Perhaps this tendency toward reactive versus proactive decision-making is why we might take a jaded view of the Great Dying and see it more as a historical curiosity than an existential lesson. We might say to ourselves, *"How interesting,"* instead of *"Holy crap!"*

The end of the Permian serves as a grim reminder to life in our biosphere as to how bad it can get. The planet became virtually uninhabitable for life. Food chains collapsed, and environmental conditions changed too quickly for most species to survive.

It might behoove us to consider what happened at the end of the Permian to cause such a catastrophe and compare that information to conditions today in the Anthropocene. Rapid climate change carries with it the threat of extinction. Eight billion Homo sapiens occupy our planet today, and we expect another two billion more by 2055. Proactive planning does not guarantee success, but not planning guarantees disaster. We can no longer afford to approach our future through a purely reactive lens. We need to do better than that.

More Oxygen Please

The Great Dying nominally occurred 252 million years ago, but debates around the length of the extinction event have produced a range of estimates from 15 million years to 200,000 years. I favor the shorter range since evolution seems to find a way to adapt over long periods, and evolution clearly didn't adapt during the Permian mass extinction.

It is also possible that harsh conditions weakened the biosphere over a longer period, and rapidly changing conditions at the end of the Permian delivered a coup de grâce.

One of the revealing facts about this mass extinction is the remarkable difference between 95 percent species loss in the oceans and 70 percent species loss on land. The large marine die-off leads some researchers to believe the underlying root cause of this disaster was a lack of oxygen.

We know from our previous discussion in this story of life how oxygen is intimately tied to life on this planet. Low oceanic oxygen levels would have led to suffocation for many species.

Warm water holds less oxygen, and the climate was warm in the Late Permian. Estimates of seawater temperatures at the end of the Permian go as high as 32 degrees C (90 degrees F).

Interestingly, the paleontological data indicate species extinction was greater in higher latitude locations than at the equator, the warmest place on the planet. The theory is that species living in normally warm waters are more adept at surviving under low and variable oxygen conditions. They evolved to deal with the problem.

However, species accustomed to cooler conditions with fewer temperature variations were more sensitive to changes in oxygen levels and unable to adjust to rapidly increasing water temperatures.

Cause and Effect

Oceans are an integral part of the planet's climate control system. They offer a huge heat reservoir to buffer atmospheric temperature changes. They are the heat pumps of the planet, cooling the atmosphere when it overheats and heating the air when more frigid conditions set in.

If the oceans overheated at the end of the Permian Period, the most likely culprit was a superheated atmosphere. Various studies have estimated average global temperatures at the end of the Permian were 20 to 30 degrees Celsius higher than today. The rapid onset of high temperatures would have triggered extreme ocean warming.

However, there was more going on at the end of the Permian Period, and geology came into play. Located in the northern regions of Siberia are the remnants of a massive outpouring of flood basalts called the Siberian Traps. Flood basalts occur when Earth's crust splits open along massive fissures, and the molten core of an underlying mantle plume (hot spot) disgorges onto the planet's surface. Rivers of lava flow like water, filling in valleys with lakes of molten rock and traversing the topography like a flood from hell. Individual flows travel hundreds of miles before freezing into solid rock.

The Siberian flood basalts covered an area estimated at seven million square kilometers (2.7 million square miles). Over one million cubic miles of rock erupted onto Earth's surface. However, the basalts were not the real threat to the global biosphere. The degassing of the mantle plume pumped unimaginable volumes of carbon dioxide and other greenhouse gases into the atmosphere, triggering rapid global warming and mass extinction.

We may want to ponder the comparison between now and then since the math is simple. The Siberian traps erupted for two million years and raised global temperatures by about 30 degrees Celsius—a rate of 1 degree every 67,000 years. Human activities during the Anthropocene have increased the global temperature by one degree Celsius in 100 years.

The Mesozoic Era: Age of the Dinosaurs

Life is tenacious, clinging to the skin of our planet against all odds.

Evolution is the fundamental force driving our biosphere into the future.

Humanity is another step on the path. Our destiny is to change, and if we cannot, we will perish.

A New Era

Geological history drew a line in the sand 252 million years ago. The Permian Period ended, and the Triassic began. But it also marked the end of the Paleozoic Era and the beginning of the Mesozoic. As a society, we find the Mesozoic Era much more interesting than its predecessor. To my knowledge, there are no popular movies about Cambrian, Ordovician, Silurian, Devonian, Carboniferous, or Permian theme parks. However, it seems we get a new movie or book about Mesozoic attractions every few months, such as Jurassic Park or Jurassic World. There is something about dinosaurs we find quite thrilling.

The Mesozoic era was the '*Age of Dinosaurs.*' The era spanned 186 million years and encompassed the Triassic, Jurassic, and Cretaceous periods. It ended in a catastrophic big bang when the Chicxulub asteroid struck the Yucatán Peninsula in Mexico. Earth basked in warmth for most of the Mesozoic, and the planet was largely ice-free, with the exception of some minor glaciation events in the Early Cretaceous.

All dark clouds have a silver lining, and the mass extinction at the end of the Permian was no exception. The good news was that the fittest survived, and when the planet returned to a more livable state, life had a clean slate on which to work its magic and embark on a new beginning. Evolution had the opportunity to do what it does best: adapt and create.

The Triassic began in unusual circumstances with an underpopulated planet. Amongst the survivors of the "Great Dying" were the archosauriforms. They continued to evolve during the Triassic, eventually ushering in the reign of the dinosaurs.

The earliest dinosaurs (Nyasasaurus parringtoni) appeared about 243 million years ago in the Middle Triassic during the Anisian age. However, they were comparatively small creatures, and another 30 million years passed before a giant species emerged. Weighing in at 3 tons and up to 27 feet long, the Plateosaurus was a forerunner of the Jurassic monsters we love to see on the big screen.

New species inevitably arise after the biosphere collapses. So we can rest safely knowing that the Anthropocene extinction is unlikely to be the end of all life. Perhaps Homo sapiens will be around after the party ends to help get the ball rolling again, or maybe not.

Carnian Pluvial Episode

The Triassic period lasted approximately 50 million years, but it was not all smooth sailing. The period of the Upper Triassic from 237 to 227 million years ago is known as the Carnian. A two million-year stretch of the Carnian from 234 to 232 million years ago is known as the Carnian Pluvial Episode. It is famous for intense climate change, where global warming pushed temperatures up by over 4 degrees Celsius, creating a very wet planet.

Pluvial periods are characterized by a wet climate with intense rainfall and widespread flooding. Researchers have described the Carnian Pluvial Episode as "*one of the most significant intervals of the past 250 million years. Within the space of two million years, the world's biota underwent major changes with dinosaurs becoming the notable incumbents.*"

Heat, humidity, and intense storms altered life on land, and the increased sedimentation from constant flooding led to the erosion of coastal marine reefs and ecosystems. Climate change was in full swing, and species extinction rates climbed. The most likely culprit for global warming during the Carnian was massive quantities of carbon dioxide and other greenhouse gases emitted into the atmosphere from volcanic eruptions. Carbon isotope analysis of rocks from the Carnian Pluvial Episode supports this hypothesis. The obvious question is, where did the massive volcanic eruptions occur?

The Wrangellia Large Igneous Province

Sometimes, geological lingo becomes a bit obtuse, but in this case, a large igneous province is just what it says. It's not just one volcano or a single fissure with magma flowing out, but a large geographical region with hundreds of volcanic features actively spewing molten lava over Earth's surface.

Today a visitor traveling the length of Canada's West Coast traverses mountain ranges from Vancouver, Canada, to Alaska. They might not realize they are traveling through the remains of an ancient geological province known as the Wrangellia Large Igneous Province. The rocks forming today's mountains are part of a plateau formed deep beneath a long-gone Triassic ocean. Magma and flood basalts erupted onto the ancient ocean floor, with accumulations reaching up to 4,000 feet in thickness.

These thick basalt flows erupted for millions of years and originally covered over half a million square miles of seafloor. As was the case with the Permian extinction, volcanic gases, not magma, served as the instigator for global climate change.

Some researchers link the Carnian Pluvial Episode with the formation of the Wrangellia basalts, but others believe the Wrangellia eruptions began at the end of the Pluvial Episode. Further work will be needed before Carnian global warming and Wrangellia volcanic emissions can be definitively tied together, but their connection is entirely plausible.

Our increasing awareness of today's Anthropocene climate change makes the Triassic story uncomfortably familiar. True, the Carnian warming was not due to industrial activity, but global warming makes no distinction between greenhouse gas sources. Whether greenhouse gases are from volcanic activity, carbon feedback cycles, or fossil fuel burning is irrelevant because the warming and environmental consequences are the same. Earth only responds to the physics of more heat trapped in the atmosphere.

End of the Mesozoic: The Last of the Dinosaurs

Approximately 55 million years ago, the Earth's average temperature surged, reaching a peak of 84 degrees Fahrenheit (29 degrees Celsius). Tropical vegetation grew in the polar regions, and sea levels were high because no ice was locked in the polar ice caps. Global temperatures were a staggering 25 degrees Fahrenheit higher than today. This peak formed the culmination of a warming trend that started in the Early Jurassic and continued building for about 150 million years. The start of this Jurassic warming trend was the beginning of one of Earth's longest episodes of continuous climatic warming. But keep in mind the "cold" starting point in the early Jurassic was still warmer than today.

The last period of the Mesozoic, the Cretaceous, ended 65 million years ago in another mass extinction. Society's romance with dinosaurs makes this end-of-Cretaceous event the most well-known of all extinction events even though it wasn't the most severe. This extinction event is also known as the K-T extinction. The term K-T denotes the geological boundary between the Cretaceous Period and the subsequent Tertiary Period. Schoolchildren and adults alike are enthralled by the story of a colossal asteroid striking the Earth and bringing the biosphere to its knees. The mighty dinosaurs disappeared, relegated to the fossil record, and this dramatic change ushered in the age of the mammals. The carnage was substantial with 75 percent of all plant and animal species disappearing. However, the event still falls at the lower end of severity when compared with previous extinctions.

The Chicxulub asteroid struck at the ancient location of the Mexican, Yucatán peninsula. The crater left by the strike is still measurable today, with a diameter of 150 kilometers (93 miles) and a depth of 20 kilometers (12 miles). Fifteen billion tons of fine-grained rock, soot, and aerosols were ejected into the atmosphere. Sunlight was blocked out, and the Earth plunged into darkness for years. The impact triggered earthquakes, volcanism, and tsunamis. Today, in the geological record, a clay layer rich in the metal iridium is found worldwide at the Cretaceous-Tertiary boundary. Iridium is a metal that is rare in the Earth's crust but prominent in asteroids.

The Deccan Traps

A less known contributor to this extinction event was the formation of the Deccan Traps in India. Flood basalts from the Deccan Traps began flowing 65 million years ago at the end of the Cretaceous Period and continued erupting for 30,000 years. These eruptions resulted in the extrusion of about 600,000 cubic miles of liquid rock onto the earth's surface, enough to cover Texas, California, and Montana. It was, of course, the volcanic gases that took their toll on the environment.

In April of 1815, Mount Tambora in Indonesia erupted, resulting in the deaths of 92,000 people. Ash from the explosion affected weather for the next year, and 1816 was the year without a summer due to temporary global cooling.

If one explosion affects global weather for a year, imagine what 30,000 years of continuous massive eruptions would do. Billions of tons of greenhouse gases would flood the atmosphere creating a global warming effect. Also other metals and toxins would enter into the biosphere. Analysis of fossil shells from 66 million years ago shows that significant global increases in mercury occurred during the formation of the Deccan Traps.

Thirty millennia of global warming, acid rain, ocean acidification, and toxic metal poisoning would have taken their toll on the biosphere before an unwelcome visitor from outer space rammed into the Yucatán Peninsula.

Sputtering remnants from the mantle plume responsible for the

Deccan Traps still pump lava through one of the most active volcanoes on the planet, Réunion, a French island in the western Indian Ocean. Despite its hyperactive eruption schedule, the island's volcanoes are insignificant compared to the glory days of the Deccan Traps.

Recent work on the mantle plume responsible for the Deccan Traps shows it to be tree-like. Approximately 1,500 kilometers below Earth's surface lies the central trunk. Sprouting from the trunk are three branches, each extending towards the surface and feeding various Indo-Austral hotspots.

The reality is that the Chicxulub asteroid was the coup de grâce for an already weakened biosphere. The short-term effects of the asteroid combined with the longer-term degassing of the Deccan Traps sought to overwhelm life on Earth. But life survived. Evolution could not be quelled. The dinosaurs departed and new species arose to fill the vacant ecological niches.

The story of life does not end with the Cretaceous, but events since then are relatively well known and thus not part of our Vanished Origins saga, a 4.5 billion year story of evolution and the challenges that our biosphere faced and overcame against all odds.

....

If there is a takeaway from life's four billion-year story, it might be that rapid climate change poses a threat to the biosphere and, hence, a threat to all life. Earth's climate is always in a state of flux, but it is primarily during those rare intervals of dramatic and rapid climate change that mass extinctions occur. The rate of environmental change overwhelms the ability of life to evolve and adapt. I would like to leave you with some thoughts on Anthropocene climate change in the following epilogue.

Epilogue

The past is a collection of imperfect memories and the future is yet to be.

The only reality is the present.

But it is in the present we can imagine and shape the future.

The story of our Vanished Origins is humbling in many ways. We are not masters of the Earth but merely the latest iteration in evolution's four-billion-year saga. Each of us will take away a unique existential message from placing our short lives in the context of life's long journey on Earth. However, there is a common context we can all engage with regarding the nature of life, the biosphere, and the shared common thread in Earth's history that connects the times when all life came under threat.

Evolution is not magic. It is a process that takes time. When environmental conditions change more rapidly than evolution can adapt and respond, mass extinctions occur. It is not the change that dooms species to extinction; rather, it is the rate of change. The rate of

Anthropocene climate change is far greater than at any other time in Earth's long history, with the exception of asteroid strikes.

Prior to the Anthropocene, one of the most rapid temperature rises recorded in Earth's history occurred 55 million years ago, as massive amounts of greenhouse gases flooded the Earth's atmosphere, causing rapid global warming. The time in question was a transitional period between the Paleocene Epoch and the Eocene Epoch, giving rise to the name Paleocene-Eocene Thermal Maximum, or PETM.

Average global temperatures reached as high as 27 degrees Celsius, which is 13 degrees higher than the 2022 average global temperature. The rate of global temperature rise leading to the PETM was extremely high from a geological perspective but still about 25 times less than the rate of the Anthropogenic temperature rise we see today. The PETM temperature spike saw the planet warm by 5 degrees Celsius (9 degrees Fahrenheit) over about 20,000 years. The sea level was 300 feet higher than it is today due to the melting of ice and the thermal expansion of seawater.

The geological data are clear that atmospheric greenhouse gases increased dramatically during the PETM, and Earth's oceans experienced rapid acidification. In total, an estimated 10 trillion tons of carbon entered into the atmosphere to create the PETM. However, the causes of that atmospheric carbon increase are speculative. Some theories suggest that massive releases of seabottom methane were responsible for the increase in atmospheric greenhouse gases. In contrast, other theories attribute the rise to the inception of the Icelandic hotspot and the associated release of CO2 from volcanic activity.

The PETM is the closest historical global warming analog that we have to today. The ocean pH fell by 0.4 units during the PETM event. It doesn't seem like much, but it represents a 100% increase in acidity from the pre-PETM levels. A pH decrease of greater than 0.1 units in human blood can kill a person. PETM changes occurred over 20,000 years at a rate of about 0.1 unit decrease in pH every 5,000 years. At the same time, atmospheric temperatures rose about 1 degree Celsius every 5000 years.

Present-day climate warming has also led to ocean acidification. Ocean pH has decreased by 0.1 units over approximately 200 years, with about 70% of that drop occurring in the last 100 years. Earth's average temperature has increased by about one degree Celsius over the same time frame. These rates of change are about 25 times greater than during the PETM event. So, even though the PETM is a reasonable analog to help understand some aspects of rapid climate warming, it cannot provide an exact match to the unheralded rate of temperature change that Earth's climate is currently undergoing

The PETM serves as a reminder that rapid global warming is possible under natural conditions. This observation applies regardless of whether the source of the carbon was volcanic-derived CO2 or methane from seabottom clathrates. Recent research into the volcanic origin of the PETM carbon suggests that up to a gigaton of carbon was released into the atmosphere each year for 10,000 to 20,000 years. Remember that global fossil fuel emissions in 2018 were approximately 37 gigatons.

The magnitude of the difference in greenhouse gas emissions between the PETM and now is in line with the observation that the rate of global yearly temperature increase today is approximately 25 times greater than the temperature rises during the PETM. This connection between past and present certainly provides food for thought.

The existential threat posed by Anthropocene climate change is apparent, and it begs the question of why humanity can't adequately respond. The science is clear, and the solutions are identifiable, yet we drift on, oblivious to the threat. I would like to finish my story with some thoughts about humans and climate change.

Wired for Survival

Humans possess genetic wiring attuned to survival. Before our distant ancestors' ascendancy over the natural environment, daily survival was the key theme of life. Humans adapted themselves to the seasons because hunting, and later farming, depended on planning, lest they starve during the winters. Less energy was devoted to wondering what would happen in 100 years because surviving the day, and then the upcoming year, was a full-time challenge. Our human instincts focus on responding to the near-term, so how do we recognize an existential threat that is centuries away.

Anthropogenic climate change poses such a threat, yet much of the world either turns away and denies the unfolding disaster or passively accepts it but doesn't act. However, the real threat is not climate change but apathy. Our inability to collectively work together and find solutions is a barrier to balancing today's needs with the welfare of future generations.

Our wiring is for daily survival, not necessarily for the survival of the species. Yes, we are biologically predisposed to reproduce, so the species survives. But if population growth and associated resource demands destroy the ecosystems where we live, we don't necessarily have the genetic circuitry that says, "Stop." Even though we can see the threat, there is a reluctance to act because the fallout from climate change will unfold on a time scale that exceeds our individual lives. We continue accumulating resources and wealth because profiting today at the expense of the future environment is a reasonable proposition for many.

More is Better

Human wiring also focuses on security. Our ancestors found that more food stored over the winter staves off starvation. Stronger walls keep out unwanted strangers. Both of these things bring increased safety and are thus desirable. If horses are essential to your society, then owning a herd of horses provides you with extra security.

Today's struggle for greater wealth is an extension of this need for security. Wealth equates to security, and we are wired to believe more is better. Thus, we eventually start accumulating wealth for its own sake, not for security. A harsh way to express this is that greed is an integral part of the human condition. But where does this lead us when populations explode, and everyone is on the path toward greater wealth? Only when the link between wealth and security disappears can we focus our full attention on the future.

Cheap, accessible energy has provided giant leaps in lifestyle and security for many, but not all, over the last 100 years, and the global standard of living has advanced over the century. However, cheap energy also flooded the skies with greenhouse gases, kick-starting climate change and global warming. It would be unfair for me to point the finger at big industry alone and say, "*Look at what they have done.*" Consumers like me also don't want to give up cheap energy.

Energy Access and Prosperity

Improvements in average living standards over the past century in countries like the USA resulted mainly from cheap and plentiful energy. Coal, oil, and natural gas were the primary energy building blocks. Accessible energy drives commerce and manufacturing, thus creating jobs and wealth. This process is a standard model playing out today in many underdeveloped countries.

However, cheap energy and innovation have done more than provide money; they have also made that money go further. When consumers buy a cheap plastic mixing bowl from Walmart, they pay their share of a production chain that begins with extracting oil from the ground as a raw material. Factories then transform the oil into plastic feedstock and send it to another plant that produces mixing bowls. Then trucks, burning fossil fuels, transport the finished bowls to a store where consumers buy them.

Virtually every item in our stores has a carbon footprint. Therefore, the more we buy, the larger the carbon footprint. The relationship between wealth, energy consumption, and carbon emissions is evident. Larger homes, more cars, frequent air travel, and rampant consumption all cost money and increase our carbon footprints.

Wealth and Climate Change

Wealth is intimately tied to resource use. As average living standards rise, the demand for more resources inevitably follows. At some point, increasing wealth encounters resource limitations, as the human desire for "more" is constrained by available resources. Mitigating climate change is about reducing carbon footprints, which will prove challenging because most people will need to live with less.

A cleaner environment only comes at a cost. Movement towards clean, carbon-free energy incurs costs, which can stress people's finances and security. Consumers may lose enthusiasm for reducing emissions if it means a decline in wealth or a rise in costs. Remember that, against a background of flat income, rising prices result in a decrease in wealth. Less wealth represents a move towards less security in most of the world. Food, shelter, healthcare, and education security are the four basics of human survival. Once those aspects of a person's life are stable, they can focus on larger issues, and there is less need to accumulate excessive wealth. The degree to which countries provide their citizens with the basics varies greatly. Infant and maternal mortality rates in the USA are far higher than in its peer countries, even though Americans spend twice as much on healthcare as the next closest country. Many Americans lack healthcare security.

We desire a clean, livable environment, but we also want to feel secure. The tension between these two aspirations creates a quandary because we must sacrifice wealth today, so future generations enjoy an environmentally healthy tomorrow. However, if sacrificing wealth means sacrificing security, we have little hope of achieving the goal of slowing the rate of climate change.

The Cost of Ignoring Climate Change

Geological history reminds us that Earth's changing environments result from a complex interplay of interactions between the atmosphere, hydrosphere, geosphere, and biosphere. Climate is, and always has been, a dynamic feature of these interactions. Put another way, the climate is continually changing. Why, then, should we worry about climate change?

Life has flourished on the Earth in one form or another for over four billion years. However, scientists estimate that approximately 99.9 percent of all species that have existed throughout history are now extinct. Organisms, including Homo sapiens, don't exist separate from the ecosystems they call home. Destroy or change the ecosystem, then the species living there must either adapt or perish.

Mass extinction events throughout geological history have one thing in common: rapid environmental change, which is often linked to rapid climate change. Sometimes, Earth has warmed too rapidly, and at other times, it cooled too quickly. But it was the rate of change, not change itself, that sent countless species into the void of extinction. Over 90 percent of Earth's species disappeared during the late Permian "*Great*

Dying" as temperatures soared and the oceans became uninhabitable.

Rate of Change

The Earth has undergone numerous cycles of warming and cooling over the past 4.5 billion years. However, what makes modern Anthropocene climate history distinct is the rate of change. At the end of the last ice age, average global temperatures experienced a geologically rapid increase of 5 °C in 6000 years (a rate of 0.83 °C per 1000 years). However, over the past 50 years, the rate of temperature increase has been equivalent to 13 °C per 1000 years. This rate of increase is unprecedented in geological history; we are into new and unexplored territory.

Life is tenacious but not invincible. Historically, species have disappeared at a rate of about 0.01 percent every 100 years. But almost five hundred species went extinct in the past decade. This rate is 50 times greater than the background level. Biological evolution needs many generations to work its magic.

The rate of change is all-important. All mass extinctions share a common theme: Climate change occurs faster than evolutionary change can keep up. Climate change and species extinction today, in the Anthropocene, are occurring more rapidly than at any other period in the Earth's long history. The cost of ignoring climate change is high.

Solutions

The search for a path forward brings to mind an oil quote from Wallace E. Pratt, "*Where oil is first found is in the minds of men.*" Oil lurks far below the ground, and geologists sift through data to tease out hints of where it may be located. When someone puts the various jig-saw pieces together and finds oil, they envision the oil field in their mind long before its discovery by the drill bit. This quote also extends to complex problem-solving in general. Viable solutions to humanity's problems will emerge from the vision and creativity of individuals as they confront climate change. Whether society implements those solutions is a different question. We also need to be realistic in our expectations.

Claiming we can stop climate change by transitioning to net-zero emissions is disingenuous. Net-zero provides a target where carbon emitted into the atmosphere is balanced out by carbon taken up in various types of carbon sinks. However, when humanity achieves its net-zero goal, the global temperature will continue to rise due to climate lag and reductions in atmospheric aerosols. Global temperatures are expected to increase by up to 1 °C over the next 40 years after achieving global net-zero emissions. So, the current Paris Agreement target of holding warming to 2 °C is actually closer to 3 °C.

This observation is not to say that reductions in fossil fuel emissions aren't necessary. These reductions are a vital part of any plan moving forward. The salient message is that a hotter planet is on the way. The critical question is, what will humanity do to mitigate the geographical, environmental, and social dislocations that climate change brings? Humans are very skilled at adaptation, and those skills must be part of building a bridge to the future.

The year was 1967, and humankind had ventured into space. As a species, we are perpetual tourists. Photo keepsakes were a must-have once we left our earthbound domain, so in 1967, the first color photos encompassing the whole Earth arrived home courtesy of the Department of Defense Gravitational Experiment (DODGE) satellite. Digital photography was not what it is today, and the color photo was really a composite of three photos taken with red, green, and blue filters. Five years later, the Apollo 17 crew took the iconic "Blue Marble" photograph. For the first time in Earth's history, the planet could be examined in a single view. The Gaian ecosystem, which we call Earth, finally came into focus, and we saw it as our largest ecosystem.

We grapple daily with environmental and climate change issues ranging in scale from microbial changes in soils to vast islands of plastic polluting our seas. Massive forest fires, melting ice sheets, and devastating hurricanes constantly capture headlines, making it easy to overlook the interconnectedness of our world. Photos of our planet, taken from afar, remind us of the big picture — life on Earth is a single web of life. At the highest level, the Earth's surface forms a connected, planetary-scale ecosystem — one that has evolved over the course of four billion years.

Ecosystems

As a refresher, an ecosystem is a group of interdependent biotic and abiotic components that function together as a sustainable unit. Importantly, ecosystems are often nested. The Tongass temperate rain forest forms an ecological community that is intimately connected with the marine ecosystems of the Pacific Ocean. Together, these two ecosystems form a larger coastal ecosystem. The Cascade Mountains of Oregon are home to temperate rain forests on their western slopes and high deserts on their eastern side. These two distinct ecosystems connect at the hip via a high mountain range, which controls rainfall, keeping one side wet and the other dry.

Our planet's surface is an interconnected web of life — a master ecosystem with multiple smaller systems nested inside. At the smallest scale, individual ecosystems are the basic units of our environment. Therefore, environmental health in a region is measured by the health of its component ecosystems. At a fundamental level, climate change presents itself as changes in local and regional ecosystems. Higher global temperatures, for example, deplete precious moisture reserves in areas such as California, leading to longer and more intense dry seasons. Dryer, hotter conditions then set the stage for larger and more damaging wildfires.

In our efforts to combat climate change, we shouldn't lose sight of the ultimate goal, which is maintaining a sustainable planetary ecosystem where humans can survive and thrive in balance with the local ecosystems surrounding them.

Life in Flux

Life is always in a state of flux and change, so the goal is not to achieve stasis or return to the past. The objective of tackling climate change is to create a dynamic and sustainable balance between humans and our environment. We depend on healthy ecosystems for our continued survival. Destroy the microbe-laden humus in the farmer's soil, and crops will fail. If we put too many nutrients or toxins into our estuarine hatcheries, our offshore fisheries will disappear.

The greatest geographic division of ecosystems on our planet is between terrigenous and marine environments. But both of these domains are affected and connected by the single most visible driver of Anthropocene climate change: global warming. The atmosphere provides this connection through its incorporation of carbon dioxide (CO2) from fossil fuel emissions.

The accumulation of greenhouse gases in the atmosphere is directly tied to a proven rise in average global temperature over the past 100 years. Temperature is one of the primary environmental conditions that affect ecosystems on land and in the sea. As temperatures change, so do ecosystems.

Our Oceans

In our oceans, global warming does more than just increase temperatures — it also increases ocean acidity. Oceans cover about 71 percent of the surface of our planet. Where the ocean and atmosphere meet, they interact and exchange carbon dioxide (CO2). Earth's current CO2 budget dictates that when CO2 enters the atmosphere from the burning of fossil fuels, only about 45 percent of this CO2 remains in the atmosphere. Land plants use approximately 25 percent of the emitted CO2, and the oceans absorb the remaining 30 percent.

Carbonate buffering refers to the process of CO2 exchange between the atmosphere and the oceans. When CO2 enters the ocean, it reacts with water to form carbonic acid, and this causes ocean acidification.

The effects of ocean acidification start at the bottom of the food chain. Higher acidity decreases the amount of calcium carbonate available for small planktonic creatures to make their shells. Groups like pteropods are an important planktonic component at the base of the food chain. Without the plankton, the ocean ecosystem collapses. If plankton can't produce viable shells because of increasing acidity, then the species won't survive, and the ocean's fish populations dependent on those plankton will decline.

The Earth is a single planetary-scale ecosystem, and changes in one part of that system will inevitably affect the whole system to some degree. Managing climate change has many aspects, but don't lose sight of the overall prize: A healthy, sustainable biosphere.

Climate Change and Food

Understanding the impacts of climate change is an integral part of managing a sustainable future. From a geological perspective, Earth is currently in an unusually cool period, and a "*normal Earth*" would be much warmer than it is today. We also know the planet is constantly in flux, and change is the norm. Historically, mass extinctions occurred through natural processes that dramatically altered Earth's ecosystems. Today, however, the Anthropocene extinctions represent rapid environmental changes caused by a single species, Homo sapiens.

The changes wrought by humankind create existential threats to our species, and we fear those changes for good reasons. Understanding the nature of those threats requires an examination of how the relationship between people, food, and ecosystems has evolved over the past 20,000 years.

The Rise of Agriculture

Approximately 12,000 years ago, Earth exited the last Pleistocene Ice Age, where glaciers covered vast areas of the planet. Climate change was rapid and dramatic, but nomadic hunter-gatherer tribes didn't ponder climate change as a problem to be solved because they were too busy adapting and surviving. As the ecosystems around them changed, they migrated and adapted their lifestyles to new environmental conditions.

As sea levels rose, coastal communities migrated inland, and when food sources disappeared, new food sources were found. Part of the key to this adaptation was a lack of permanence. People moved to where the food source was. Migration was a key component of survival.

The first agricultural revolution changed this dynamic. With the advent of crops and domesticated animals, communities settled into fixed locations for extended periods yet still maintained sufficient food sources. This change gave birth to the age of cities and empires, where adequate food security allowed human creativity to flourish, producing new technologies. The rise and fall of empires didn't extinguish the true strength of Homo sapiens, the ability to pass knowledge on from generation to generation.

The Role of Cities

One of the keys to understanding the threats posed by climate change lies in the role cities play in our societies. We take them for granted in places like the USA and forget that 54 percent of the population was rural in 1910, but today, 81 percent of the population is urban. Traditionally, rural populations aligned closely with the ecosystems surrounding them. Food sources were local, and a community's population-carrying capacity directly reflected the amount of food an ecosystem could produce.

However, the advent of large cities broke this dependence on local ecosystems. It gave rise to agricultural systems where industrialized farms produced food products and shipped them to large urban centers. Readily available food allowed these urban centers to grow into the metropolises we enjoy today. But this process produced a system where most of our population is disconnected from the ecosystems that sustain them. Large cities are no longer supported by local ecosystems. Food is sourced from around the world and we are dependent on regional and global ecosystems. The fate of these cities is tied to the global effects of climate change.

Food Security

Climate change threatens this relationship between rural and urban settings when environmental changes jeopardize major food sources. So, one of the concerns with climate change is food security. But even this threat is nuanced. In wealthy countries, many people were unaware of the 2007–2008 global food crisis, which was triggered by high oil prices, increased biofuel production, weather shocks, and trade restrictions, resulting in grain shortages and jeopardizing food supplies. Food, like other commodities, rises in value when it is in short supply. For those with ample financial resources, the 2007–08 grain shortage resulted in household budget trade-offs, as more money was allocated to food and less to other activities. But for the poor, higher prices translated into less food. These shortages triggered social unrest, malnutrition in children, and even starvation.

If climate change causes food production to falter, then rising prices take their toll, with the poorest and most vulnerable populations being the first to feel the impact. From a global perspective, food scarcity is one of the existential threats posed by climate change. In a world where almost eight billion mouths need feeding each day, changes in the climate, which alter or diminish agricultural output, also threaten social stability and the lives of the poor. Remember, approximately 3.5 billion people, almost half the world's population, live in poverty, struggling daily to provide food and shelter. Small changes in food availability or costs have life-and-death consequences for half of the people on our planet.

Climate Change and Water

In the context of climate change, water is a bit of a Goldilocks situation, where too much or too little is disruptive; therefore, we want just the right amount in our ecosystems. A farmer might long for the rain to water his parched land, but too much water creates flooding, which washes away his crops. Even in a world where oceans and seas cover 71 percent of the planet's surface, too much seawater is a threat. Then there is the water in the ground, which supplies the needs of vast agricultural complexes. Both over-pumping and drought-reduced recharge currently impact the water supplies for America's breadbasket and other critical agricultural regions.

The Ogallala

Hidden beneath the high plains of America's mid-continent is the Ogallala aquifer. It stretches from South Dakota to West Texas and provides water to parts of eight separate states. One-sixth of the world's grain production and America's breadbasket depend on groundwater from the Ogallala aquifer. However, we ask too much of this natural wonder, which currently supplies 30 percent of all irrigation water in the United States. Because water doesn't magically appear below the ground, the aquifer must continually recharge from surface water. Unfortunately, the recharge rate for the aquifer is slower than the rate of withdrawal.

The Ogallala aquifer has stored as much water as Lake Huron (29 billion acre-feet) in the past. However, many parts of the aquifer are in decline today, resulting in falling water levels, dry wells, and the aquifer's inability to meet annual demand. Water volumes in western Kansas, for example, are down by 60 percent from original levels.

The decline of the Ogallala aquifer is only partially due to climate change. Its decline is also driven by insatiable demand and poor water management. This aquifer has historically been the backbone of a vibrant farming economy, and rural farming communities across America's breadbasket depend on its water. Increasing water demand and shrinking supplies of recharge water are challenging lifestyles and damaging local economies.

Megadrought

One of the manifestations of climate change is a two-decade drought in America's West, leaving many regions hot, dry, and thirsty. However, the situation may worsen before it improves. Tree ring data indicate that megadroughts have occurred four times in the past 1,200 years. The current drought began in 2000, and the period from 2000 to 2018 was the second driest 19-year period on record since the 800s.

While it is true that natural causes facilitate the occurrence of megadroughts, this does not mean human influences aren't also at work. Anthropocene climate change and global warming are human-induced changes that amplify the severity of this megadrought.

Global warming has increased the average global temperature by about one degree Celsius over the long-term trend. But this number averages out both larger and smaller regional trends. Average annual temperatures in the American West have increased by 1.2 degrees Celsius over the norm. Higher temperatures result in more moisture loss, so global warming works to amplify the drought severity. Researchers attribute up to 50 percent of the severity of the current drought to human activity. Climate change affects the distribution of water on our continents. There will be winners and losers as water resources become scarce in populated areas that depend on surface and groundwater supplies.

Looming climate-change-related water problems on land often play second fiddle to our most significant concern about water — rising sea levels. Under most current, realistic scenarios, the world's two major ice sheets are engaged in a slow meltdown into oblivion. The Antarctic and Greenland ice caps are feeling the effects of global warming. It is true that a full meltdown is potentially hundreds of years out, but even in the early stages, the impact on coastal cities will be significant.

Our Coastal Cities

The threat of sea-level rise is disproportionately large for humans because of the way we have historically sought to settle near the oceans. About a third of the world's population lives within 100 meters of sea level. But our return to an ice-free planet will submerge the first 66 meters of coastal land. In the USA alone, cities such as Houston, Miami, New Orleans, Charleston, New York, Boston, and Washington, D.C. will all disappear beneath the waves.

Much of New Orleans already lies below mean sea level and is susceptible to small rises in the ocean. Other cities like Miami, perched at sea level, are already feeling the pinch from increased tidal and storm flooding. But the cities at highest risk are in Asia, not the Americas. With a 4 degree Celsius global temperature rise, up to three-quarters of a billion Asians will be living below sea level. In Shanghai, China alone, over 22 million people will be displaced by encroaching seas.

We have chosen to locate many of our major cities by the sea. These decisions have their roots in trade, economics, and a traditional reliance on shipping to build wealth and power. Despite the good reasons for locating near the sea, the threat remains. Our habitats have become fixed, and our ability to migrate to happier hunting grounds, as our ancestors did, is limited. When all land is occupied or owned, where do you migrate to? The only viable answer is to either new or existing cities above sea level. Coastal farmers will fare no better. It is not easy to take a 500-acre farming business and quickly relocate somewhere else.

Colorado River

Water becomes a priority as climate changes force mass migrations. These changes needn't be just rising seas. As access to freshwater changes, the fortunes of populated areas will wax and wane. The Colorado River flows 1,450 miles from its headwaters in the Rocky Mountains to the Gulf of California, and it provides water for some 40 million people. Additionally, it provides irrigation for 5.5 million acres and has a hydroelectric power generation capacity of 4,200 megawatts. However, the water usage demands on this river now exceed its annual flow capacity.

The 20th-century annual average discharge for the Colorado River was approximately 16.5 million acre-feet (MAF), but long-term drought conditions in the 21st century have reduced annual inflow, resulting in lower river flows. However, demand for the Colorado River's water is increasing and started seriously exceeding supply at the beginning of the 21st century. In 2009, for example, the yearly water recharge supplied via precipitation was about 12 MAF, and demand was just under 15 MAF.

Major cities like Phoenix, Tucson, Las Vegas, and Los Angeles depend on water from the Colorado River. The bottom line is that future aspirations for water use appear unrealistic under current plans due to their projected water demand of between 18 and 20 MAF per year, compared to a projected annual supply of 14 to 15 MAF. Water insecurity will curtail population growth, but if conditions are severe enough, people may need to move. Such changes are difficult for those with wealth and devastating for those living on the edge of poverty.

The Future

Large human populations have become fixed in areas where climate change may affect their ability to survive. Complex relationships exist between ecosystems, food, water, and our habitats. We are right to be concerned about a future where these relationships change as the climate changes, forcing human migrations and the development of new agricultural strategies. But we shouldn't fear climate change. Instead, we should concentrate our energy on devising solutions.

What we should fear is inaction and procrastination on the part of governments, both local and federal. A strategy to maintain the status quo is a road to failure. Climate change is not an existential threat unless we collectively decide to let it be.

Made in the USA
Monee, IL
04 November 2026

67384438R00066